As a people, we are:

Ever Hearing ...

and

Ever Seeing...

A Book Reflecting on Science

Through the Ears and Eyes of a Science Teacher

David Reid

Plumbline Publishing and Software
13287 W. Montana Place
Lakewood, Colorado 80228

About the front cover:

The cover photograph was generated as part of a study partially supported by a Title III Individual Teacher Grant from the Colorado Department of Education, 1976. The report title is: "Wave Interference from a Pulsating, Elliptical Boundary." Student assistants for this study included: Paul Applegate, Richard Klamman, and Gregory Whitehair.

The wave interference patterns within the elliptical boundary suggests the foci might be 'out-of-phase.' This type of evidence led the author to the development of a wave interference model for planetary orbits in the Solar System.

Plumbline Publishing and Software Company
13287 W. Montana Place
Lakewood, Colorado 80228

Library of Congress Card Catalog Card Number: 99-93083

ISBN 0-9669041-0-9

Unless otherwise noted, all Scriptural quotations are taken from the Holy Bible: New International Version ® (NIV®). Copyright © 1973, 1978, and 1984 by International Bible Society. Another version used is the King James Version.

Printed in the United States of America

CONTENTS

Preface

How can we know truth? Science is a brewing cauldron of information: questions, hypotheses, experimental and theoretical results, and a trusted template for the interrogation of all determinate phenomena. For many practitioners of science the activity is analogous to a religious experience; yet, most recognize the tentative character of science, its formidable task of providing concise descriptions and interpretation for the awesome events within nature.

Science contemplates the whole scale of nature, from microcosm to macrocosm, and its multitude of interactions. The search for a conceptual framework (facts, principles, laws, and theories) to explain interactions at work within nature is among the major objectives of modern science. A unified view for nature would provide a level of confidence for the descriptive tools used by scientists to interrogate and interpret the unknown.

Modeling is a heartbeat within modern science. And, modeling science activities in the classroom is fundamental to quality, science instruction. Many of the ideas expressed in this book were part of the instructional strategies used to teach high school students about the joy and wonder of science. In particular, Chapter 3 develops mathematical modeling as an indispensable tool for the physical scientist and Chapter 4 provides an example of science suitable for student and/or teacher project work.

Part I of the book is reflective of the structure and language of science. Scientific inquiry has been, is, and will continue to be a compelling agent for change through the identification, analysis, and interpretation of natural events and processes. While discovering and interrogating the mysteries nature are of primary concern to scientists, the communication of findings to the scientific community and a literate public are among their great challenges.

Science can describe a system displaying particular trends and/or patterns among its operative components. It can interpret a particular interaction; it can predict the response of a system under controlled or defined conditions. Science has extended man's realm-of-experience and view of the universe well beyond the crucible of our existence.

Part II of the book is an example of the steps used to form a scientific model. It begins with a survey of evidence to support a scientific hypothesis, the structure and dynamics of satellite motion within the Solar System, inferring the Solar System to be a resonant, gravitational system. Developing and using a wave interference model, a quantum, gravitational view for the Solar System is proposed. Fundamental values associated with a quantum view of gravitation within the Solar System are hypothesized.

Lastly, one of the more important aspects of science is its ability to predict future applications and/or events. Prediction is an important part of science; and, arguments for two predictions are addressed within this book. Both predictions are based on principles of astrophysics: an imminent, extended solar minimum, and a quantum view for orbital motion in the Solar System. Whether these predictions prove to be right or wrong is important, but the quest for a meaningful response to a challenging question propels scientific literacy, a search for truth and understanding. Science has a standing invitation from nature; come: observe, investigate, contemplate, communicate, and learn.

Acknowledgements

To my wife, Marilynn, for love, encouragement, and trust in a twenty plus year gallop through uncharted regions: I love you, and I thank you. To our family: 1) Terri, Larry, Jon Paul, and Elizabeth, 2) Charles, Debi, Cassandra and Corinne, and 3) Heather, Mike, Akeeley, Dakota, and Talon I give thanks for your patience and understanding.

Thank you to: Charles, our son, who provided editorial comment; Robert Nyland who shared information about printing procedures; Ancel W. Lewis for his legal counsel on publication and copyright regulations; and **BookCrafters** (Chelsea, Michigan), printers for this book. Thank you one and all for helping to soften the rough spikes associated with this publication!

Thank you to: the Jefferson County Public Schools, colleagues in science instruction, and former students for a refiners experience. This wholesome, learning environment nurtured and cauterized many of scientific views shared in this work.

Thank you to the Physical Science Study Committee for its organization and presentation of a quality, secondary science, physics curriculum. This masterful work significantly influenced my view of science, greatly affecting the selection of instructional strategies used with students and ideas shared with colleagues.

Lastly, and most profoundly, to the Lord GOD for being who You say You are: the Creator and Sustainer of all good things. Thank You Father GOD for the opportunity to know You, to believe You, and to serve You.

David Reid

Ever   Hearing ...

And

Ever   Seeing ...

Part I: About Science

Part II: Examples of Science

1999

Common terms used within this book:

1. Nature is:
    a. a manifestation of truth, and
    b. <u>not</u> bound to respond in accord with a scientific interpretation describing its observed behavior.

2. Interpretations for Natural Events[1]
    a. subject to human limitations

    b. collection of evidence through observation and investigation
      1) utilizing instrumentation to extend human, sensorial perceptions,
      2) providing an empirical pathway to understanding and truth, or

    c. designing, building, and testing a model or a simulator,

    d. verification process: analysis, testing, refinement and/or acceptance, and

    e. communication to a scientific community and a scientifically literate public.

3. Framework for Scientific Analysis and Communication[2]:
    a. **Property / Variable** - an essential trait, feature or quality (ex: length, mass, volume, position, temperature, charge, color, texture, hardness, etc.) associated with an object at a given time;

    b. **State** - a set of properties used to characterize a condition among inter-relating variables at a given time;

    c. **System** - an isolated set of interactive objects in a descriptive state;

    d. **Rule/Fact** - fundamental quantity used to describe an object's setting (ex: geometry, standard for measurement, operational procedure, field structure, etc.); a repetitious characteristic or behavior;

    e. **Principle** - a natural tendency (ex: action/reaction, effect/cause); a fundamental coupling (ex: force/object, equilibrium/particles, symmetry, observational uncertainty, etc.);

    f. **Conservation Law** - an analytical concept used to compare the initial linear momentum, angular momentum, and family number (fundamental particles: electron, muon, and baryon); and

    g. **Theory** - a highest order of connected, scientific logic; characterizing all physical interactions and transformations.
    A theory must be:
      1) based on experimental evidence,
      2) organized in a format reflective of simplicity and generality,
      3) consistent with interpretations from related, scientific investigations, and
      4) predict an 'effect and cause' for past and/or future events.

Part I: About Science

Science is an impressive, if not compulsive, human activity. Learning about, and probing the mysteries of the universe is a bit like watching children progress through their stages of acquired motion: 1) coordinating arm and leg motion to form a 'roll,' 2) crawling about and pulling upward into a stand, 3) a taking of those first, uncertain steps; then, walking, and 4) putting multiple strides together; enabling rapid motion, running. Witnessing controlled motion is an exhilarating experience; yet, for most of us, rapid motion is periodically punctuated with an unplanned stumble, if not a fall.

Those falls can be painful, but they are always full of opportunity to learn. A key to a positive, learning experience is to get up and try again. Advancing the whole of science toward parity with truth is a challenging task. Learning from prior experiences, engaging and interacting with current challenges, and hypothesizing about the possible challenges of tomorrow have produced a crescendo of trusted information, a patchwork of momentous development.

We are immersed in a historic legacy of human effort. Today's quality of life is pillared on the experiences and trials of our ancestors. What an exciting task it is to reflect on their vision of living and their dedication to learning. Science is but one component contributing our quality of life. Yet, science is and has been a prime mover in shaping the history of mankind.

Part I provides a brief overview of science and reflects on science as a human activity.

Chapter One: A Prologue On Science

The development of science is dependent on trusted, scientific results (content) which come from a tested investigative framework (process). Chapter One introduces the parts of scientific inquiry influencing the credibility and the development of science. These components include:

- identifying a natural event or process to study, describe, and explain
  1) form an inquiry question and hypothesis
  2) gather evidence/data to analyze and interpret
  3) express the experimental finding in a mathematical model,

- communicating scientific results to the scientific community and an interested public, and

- refining and completing a trusted, conceptual framework to guide scientific inquiry.

Science views nature with deterministic 'ears and eyes,' confident that nature is capable of description and interpretation. A determinate view of nature drives the scientific inquiry process; therefore, the disciplines of science are interdependent. The interdependency of science is a form of evidence supporting the scientific search for a unified view for nature.

Chapter Two - Affecters of Science: Scientists, Institutions, and Events

Including 1600, the turn of each century has ushered forth at least one new, scientific discovery. This fresh prospect for gaining information about an awesome, but complex universe has propelled the development of science. These discoveries have brought opportunity to review the historical development of science; identifying strengths and weaknesses in its methodology and results. Both intellectual arguments and laboratory results were reviewed and critiqued. These discoveries have forged new opportunities for technological development and research science.

Among the significant factors influencing the development of science during this post-renaissance era has been the intellectual and philosophical contributions from Judeo-Christian religion. Many of these influences have affected the development of science positively, others have produced seeds of discord between science and religion.

Change is never neutral. The winds of change affect what happens to a citizenry. Orchestrated change stimulates what is thought, what is accomplished, and what remains incomplete and/or untouched.

Chapter Three - Mathematical Modeling: A Process of Synthesis in Science

Modeling is the heartbeat of modern science; and, a mathematical model is an important, scientific tool of expression and synthesis. A mathematical model expresses an interrelationship among the operational variables within a system. A mathematical model provides a window to reflect on a scientific result and predict a forthcoming event. Experiments observe, probe, and measure nature's response to a given condition. Theoreticians apply this mathematical framework to intellectually probe problems, if not, dissect the logical mooring of science and project toward the hidden mysteries within nature.

One form of scientific unification calls for a mathematical continuity among the profusion of experimental constants. Chapter Three introduces this logic and makes use of the 'kinetic-molecular model' as a historic example for creative synthesis in science.

CHAPTER ONE: A Prologue On SCIENCE

Science is an intense learning activity. Science demands an incorporation of all disciplines, drawing information and technique from the modern academics: history, art, music, linguistics, and/or mathematics. There is no best way of doing science. But, scientific inquiry always begins with a well-defined question. Science views its intellectual and experimental results with a tentative eye. Science is 'ever hearing and ever seeing,' questioning and testing its results; relentlessly probing the composition, structure, and interactions within nature.

Scientists form a unique community with goals. These goals include:

- searching nature for observational evidence about repetitive, physical events and providing description and interpretation for these events,

- communicating theoretical and/or experimental results for review and critique; the findings must be tested and either confirmed or rejected,

- providing leadership to direct the thrust of scientific research and influencing the education and training for all practitioners of science.

Both science and science education are steeped in a historical perspective. Understanding the historical synthesis and development of science is a powerful, instructional framework. The how and why associated with each step of scientific development are subject to the eye of scrutiny and introspection. This surveillance validates the inquiry process, cauterizing the intellectual and experimental logic used to design, initiate, and culminate each investigation. The training process for new scientist presents a natural review process for scientific inquiry, an enactment of the purpose and strategy for any given investigation. This historical review of science provides an opportunity to reflect on a rich legacy of scientific discovery and creatively project toward the unknown, calling for answers to new questions. The history of science, with its failures and successes, guides the scientific, inquiry process and shapes the emerging development of science.

The Inquiry Process

Scientists study repetitious, natural events that are either interesting or, could impact the quality of life within the sphere of human influence. Scientists question what they hear and what they see. They form hypotheses to respond to their questions, they gather and analyze evidence to describe and interpret these events. These practitioners of science look for patterns and trend among the evidence and results, searching for clues to connect their observations with earlier works of scientific inquiry. A physical event may have an astronomical, a terrestrial, and/or a microcosmic connection; and, the frequency and/or period for the recurrent event becomes a signal for its importance and insures its introduction to the scientific, inquiry process.

Scientific inquiry begins with the framing of a good question about an observable or a definable event and/or process. Can this event be isolated, observed, and studied in its natural setting? What probable cause can be associated with this effect? How is this event connected to other, identifiable events or processes within nature? Some interesting physical events or processes present a formidable challenge for inquiry; a hostile environment or a remote location. If investigators have limited access to an event, then designing an artificial setting (a scaled model) for scientific inquiry may be the only alternative.

Nuclear scientists study impact collisions, fission, and fusion among a profusion of invisible particles within a hostile environment. Scientists use high energy accelerators and/or confinement chambers to probe the mysteries of composition and structure within the microcosm, replicating an alien environment to observe and study interactions beyond the limits of their natural setting. The design, construction, and operation of these simulators present extraordinary challenges for engineering and scientific inquiry.

> **Designing an Inquiry Model**

Framing an explicit question about the event or process is coupled with forming of a hypothesis, a search for a feasible answer to the question. Then follows the design of the investigative process, an inquiry model. Identifying, isolating, observing, and defining system variables become the next task. The experimental boundaries and assumptions must be set. These experimental parameters control the investigation and focus the gathering of scientific information describing the event. A use of digital information to describe the state of a system has become a powerful hallmark of modern science. Measured, numerical values for a given quantity at a specific time form invaluable information to identify trends within a system. Change for a variable and/or the change among two or more variables can be monitored, analyzed, and expressed in a mathematical model.

A bolt of lightning is an awesome event to witness and explore. It may announce a coming storm. It may cause severe damage to a living and/or an inanimate system. Being in a right place at a right time to study lightning is one variable among many experimental variables. Knowing lightning strikes have a greater probability on a mountain peak rather then on a flat terrain is helpful. It narrows the search to locate a likely site to study this event. Then, comes an observing of repetitive events, defining the lightning system with its variables, and testing the experimental question. Is the question believable? Is the hypothesis plausible? Does the inquiry model satisfy the conditions implied by the question, hypothesis, and the investigative assumptions?

Hypothesizing an outcome for a given action can contribute to the ultimate success of a given inquiry. For example, positioning a lightning rod on the mountain peak and connecting the rod to scientific equipment designed to measure a proper range of electrical values for a lightning strike can affect the result of an investigation. Is the rod positioned correctly? What conditions precede an electrical discharge of lightning? Will the electrical instrumentation be adequate for the range of data describing the electric potential and the discharge current? Is there a relationship between the electric potential just prior to a lightning discharge and the current associated with a lightning strike? Are the conditions associated with a discharge of lightning analogous to the conditions associated with an electric discharge (a spark) in a laboratory setting?

These tangential questions tend to clarify and focus the inquiry process. They tend to move an investigative strategy from a system view to a component or part view. This whole-to-piece simplification is an example of **deductive logic** and it is a powerful way for scientists to view a complex event. As the function and operation for each component within the system is understood, its relationship with the other, system components can be explored and described.

> Observational Evidence

The acquisition of reliable data is the key component for any scientific, inquiry model. Data may come from any of several sources, including data from an earlier study, a blending of data from an earlier study and newly collected data, or it may require a strategy to collect new data. A literature search, looking for an experimental connection between the current experimental design and a result from an earlier study, can prove helpful. Connecting the experimental result with a prior investigation adds credibility for both the earlier work and the current inquiry.

The strategy used to collect data is fundamental to the acceptance of any experimental result. The data must reflect a connection between the experimental question and its allied hypothesis. The thoughtful orderly strategy for data acquisition is mirrored through in its logical display. Both the strategy for the collection of data and the organization of the data during an analysis can affect the quality of an interpretation.

The subtle uncertainties associated with observation, measurement, and/or the description given for an interaction among particles can bring a bias to the event being observed. Getting precise information to describe the event's location, size, time, and other notable characteristics becomes a formidable task. For investigations within the microcosm, the Heisenberg uncertainty principle predicts an observer's inability to obtain exact information about a given particle or group of particles at an instant of time. The observer's presence is a disruption for the event; also, the traverse time between successive signals from the event will generate another 'blip' in the information continuity. Most observational ambiguities are readily smoothed by the techniques of statistical mathematics. Yet, this gap in a 'continuous flow' of information present a philosophical concern for some practitioners of science; and, is of little or no concern to other scientists. Within the microcosm, observational certainty seems to reflect a limitation, a natural consequence of viewing a microsystem. Who said science is tentative?

> Mathematical Models

The analysis of observational evidence is often summarized with a mathematical model. This explicit expression defines an operational pattern among the system variables. If the system has three or more variables, then a strategy to control one or more of the variables (parameters) must be considered. Controlling system parameters, those quantities capable of systematic variation, can simplify the identification of a pattern and/or trend among the remaining variables. Graphical comparisons often lead the inquiry toward a culmination; providing evidence for a mathematical interpretation describing the interrelating system variables.

As each system variable contributes its expressed effect to the operation of the system, the cumulative, operational expression for all of the system variables is noted with the emerging mathematical model. The isolation of system variables and system parameters is a deductive process. The description of a system becomes more comprehensive with the definition and quantification of each variable (an **inductive logic**), enhancing the system's mathematical model. Inductive and deductive logic are reversible processes usually working in tandem within an inquiry process to structure and communicate a plausible result.

Within operational parameters (limits and/or controlled conditions) for a system, a mathematical model provides a platform to calculate and/or predict systemic response: past, present, and future. Predicting the response of a system to a given set of conditions becomes an avenue to ask new questions about the function and operation of the system. Could system variables and/or system parameters be modified to establish a more efficiently operating system? How will a system respond to a modification of its scale or dimensional size?

Science is always testing its description and interpretation of an event or the operation of a system. Have all of the variables within the system been identified? Was a system parameter held implicitly constant during a given investigation? The continual testing of the system, searching for minute variations, may provide the evidence to further separate a traditional, experimental constant into new components. Each refinement in a mathematical model brings a greater degree of reliability to describe the operation of the system and predict its response to given set of conditions.

A scientific result expressed in a mathematical model is eloquently simple and succinct in form. A good mathematical model is reflective of a credible inquiry model and lends credence to the resultant interpretation.

The successful application of new technology for the home, the marketplace, and the research and development community depends on the predictability of the product to perform under specific conditions. Testing and predicting the response for a system and/or product clearly brings an element of confidence, purpose, and/or trust to the scientific, inquiry process. Predicting the response of a system with a set of conditions offers system operators an opportunity to optimize its use.

Validation and a Conceptual Framework
The review and critique of a scientific result is a form of product evaluation or testing. The confirmation of an experimental or theoretical finding brings both acceptance of the result and a verification of the conceptual framework. The successful application of the scientific, conceptual framework to solve a problem or to provide a reasonable answer for a question about nature enhances the credibility of this scientific tool.

Each part of the scientific, conceptual framework is based on experimental results, outcomes demonstrating a common 'trend or pattern' among a multitude of experimental results. The operational components for the conceptual framework include: facts, rules, principles, conservation laws, and/or theories. This framework guides a scientist's logic at each step of the inquiry process.

Historically, the application and verification of the conceptual framework greatly affect the way scientists view the universe. Each operational component of the framework has been tested and scrutinized prior to its acceptance by the scientific community. Scientists understand the assumption of truth brings a bias to the inquiry process, affecting the totality of science. Each step within an inquiry model, from the framing of a question through the communication of a result or finding, depends on the validity of scientific facts, rules, principles, conservation laws, and/or theories. Practitioners of science move through nature's maze of subtle signatures, led by these intellectual and experiential guideposts. Scientific interrogation demands the use of a valid conceptual framework and its successful application to solve specific problems by skilled investigators.

All scientific results are subject to persistent testing and/or surveillance. Reviewing the logic and strategy for an investigative model has become a hallmark of science, particularly focusing on the experimental data: its analysis, its interpretation, and its connection to earlier experimental results. Modifying historical investigations, perhaps looking for ancillary results or gaps in logic, can tighten the informational tapestry of science.

If a flaw in the inquiry process is identified, then each step in the inquiry process must be examined. The original question, the hypothesis, the investigative assumptions, and the experimental procedures must submit to this scrutinizing process. Checking for a proper use of the conceptual framework is mandatory. Does each step reflect an appropriate use of logic? Should any modification to an inquiry model be proposed, it must insure a consistency among the inquiry components: the question, the hypotheses, the assumptions, the acquisition and analysis of data, and the experimental result.

A harmony among the components of scientific inquiry: question, hypothesis, assumptions, and experimental evidence, signal a readiness to proceed with the culmination of an investigation. Analyzing the data and preparing a mathematical model to describe the interrelationship for the system variables govern the interpretation for a given event or process. Communication about a finding or result must be concise, both simple and precise.

> Communication of Results

A forum for the open exchange of information is axiomatic to science. Critiquing and repeating an earlier investigation can present optional interpretations for a given result. A different interpretation may stimulate a range of new ideas and inquiry designs. An exchange of differing views on an interpretation of an experimental result or a radical, scientific issue is vital to a search for truth. Scientific literacy is a high priority for science educators, scientists, and an interested public. All of these groups depend on leadership and guidance from the scientific community to guide the credible development of science.

Scientific publications provide an excellent, historical log for scientific inquiry and views on scientific issues. These publications document culminary expressions about the development of science. Scientific publications are an important, sifting agent; hence, they are reflective of the history of science. They tend to be cautious in guiding innovative flare (if not fluff) and/or issues affecting projected development. Scientific publications carefully review all published articles, forming a foundation of legitimacy for verifiable results and a platform of moderation for debatable issues. They document both affirming and dissenting views on theoretical and experimental results and controversial, scientific issues.

Periodic meetings and/or conventions of scientific societies focus the mission of a particular sector within the scientific community. Current discoveries, new methodologies, mounting concerns and/or hopes can be presented and discussed with civility. New and creative views of nature can be openly discussed and debated. These gatherings are important for personal and corporate communication about science, a form of science training and science education.

Determinacy, Interdependency, and Unification

The late Renaissance ushered in new ways of looking at almost everything. Man's perceived position within creation submitted to intense questioning. This new mode of inquiry invigorated science and scientific investigation. By the 17th century curious practitioners of this new mode of thought and inquiry began noting patterns of simplicity and tenets of order within nature. The rise of scientific inquiry, its new views of the universe, its newfound ability to identify, observe, describe, and predict trends for natural phenomena established an awesome platform of perception for mankind. Early scientific inquiries inferred a determinate nature; a nature that is comprehensible, capable of description and explanation.

A determinate view of our universe is parsimonic, if not simplistic. This does not mean the structures and mechanical operations within the universe are simple; rather, science tends to view the universe deductively, as a collection of independent systems. The identification and description of the known systems in the universe is a primary objective for science. Science wants to provide an accurate description for each system: its composition, its structure, and its function in nature. This is an awesome challenge and responsibility. With successful explanations for natural phenomena came another trend, an insight into similar operations occurring in vastly different systems. **If** different systems, perhaps vastly different in scale and function, exhibit analogous characteristics, **then** defining a connecting pathway between these systems may be within the realm of possibility.

If connecting pathway between systems can be identified, **then** a unification of science becomes feasible. This inductive view of the universe sees an ultimate, physical system; one filled with smaller systems each capable of description and interpretation, and harmoniously dependent on one another.

An inductive view of the universe presents science with a possibility for a grand unification among all systems. The vast difference in scale, operation, and function for each of these systems present science with a formidable challenge. A unification strategy to describe the operation and function for all systems within the chemical, biological, terrestrial, and astronomical realms would be a triumphant linkage for science, its conceptual framework and its quest for truth.

Isaac Newton[3] was a compelling scientist. His view of the universe came early in the development of modern science and his work continues to impact the current development of science. In Newton's day scientific inquiry was in its infancy and consistent information describing the events of nature was lacking. A conceptual framework to guide scientific thought and inquiry did not exist. The testing of ideas was vulnerable to error and misinterpretation. Good observational evidence was rare; yet, the contributions of practitioners like Newton, Nicholas Copernicus, Tycho Brahe, Johannes Kepler[4], Galileo Galilei[5], Christiaan Huygens and many others provided science with an excellent beginning. Persistent curiosity and methodical investigations established scientific inquiry as a powerful tool to question and explain the cause and effect relationships identified throughout nature.

Early definitions and investigations of the natural realm presented only shadows of complete explanation. Results from these early probes of nature detected incredible patterns; geometric structures of inexorable beauty and exquisite order. Again, these patterns encouraged the forging of a philosophical assumption about nature; namely, its composition, structure, and harmonious interrelationship are determinate. The successes associated with the interpretation of natural phenomena and the predictive character of these results began moving science from a stage of acknowledged mysticism toward a level of designing and implementing systematic inquiries which produced reasonable description and explanation for identifiable events in nature.

The early quantitative developments in science were dependent on conjecture and inference. The boundaries for assumption and hypothesis fell short because of a lack of observational evidence and experimental results. The inquiry process was subject to repeated modification, trial-and-error. This forage of logic was fertile ground for misinterpretation. Slowly, repetitive observations, consistent experimental results, and predictable patterns among scientific investigations began an illumination of the mooring for a conceptual framework. This framework has become a 'Rosetta stone' for scientific inquiry, theoretical and experimental guideposts for the interrogation of nature.

During the 18th century Newton's particle model for light was highly regarded by most members of the scientific community. The particle-like behavior of light with its invisible components began to affect scientific thought about the possible composition and structure of the universe. This bias was influential on other manifestations of energy particularly, an initial view of thermal energy and its impact on system variables. The successes of the particle model seemed boundless; yet, by early 19th century the discovery of interference events with light and a measurement for the speed of light in various, transparent media were among the causes for scientists to look for another view of nature other than Newton's particle model.

Although Huygens was a contemporary of Newton, his wave model for light had only a modest following as the 19th century began. The identification of interference events with light reinforced a wave like view for light. Slowly, but steadily, the particle model for light gave way to a more comprehensive wave model to describe the nature of light. And, the wave model began to influence hypotheses about other forms of energy propagation. This allegiance to a 'wave view' continued until late in the 19th century when a series of experimental results, particularly the results of the photoelectric investigation and black body radiation studies, began signaling a recurrent particle behavior for energy transitions and transformations within the microcosm.

Arguments about the nature of light, a 'wave model' or a 'particle model,' stimulated a search for a unifying view for nature. Unification logic was being further advanced by a clarification of the conceptual framework, particularly the development of the conservation laws. The conservation of mass, linear and angular momentum, electric charge, and eventually the conservation of energy were connecting scientific results on an unprecedented scale. By the end of the 19th century the scientific, conceptual framework was still evolving, but its scientific function as a tool guiding the inquiry process was apparent. While some within science remained tentative about the use of this conceptual tool, most scientists accepted the linear development of scientific facts, rules and principles, conservation laws, and acknowledged theories. Because a theory reflects the highest order of connection to experiential science, a theory became a point of interest, concern, and testing for an engaging cadre of modern scientists.

A scientific theory is an example of an intertwining of experimental and/or intellectual details. Theories may predict investigations yet to be tested, but a continuous trail of evidence and experimental results must be the foundation for its expression. Models differ from theories in their allowance for the existence of information gaps. Identifying, exploring, and explaining the gaps in physical evidence describing nature help complete science; also, it refines and clarifies the conceptual framework. When informational gaps give way to a continuous path of evidence and explanation, scientific models are forged into a theory.

A theory brings a powerful tool of reflection and prediction to scientific inquiry. The acceptance of a theory is not a guarantee of a lasting position on the conceptual framework. Review and testing continue to attack its simplicity and its generality. One negative, experimental result will move a theory back in status to a model, if not deletion. Additional experimental evidence must be collected before a new theory can be presented to replace the errant theory. A new theory must account for historical, current, and future experimental results. The 'correspondence principle' is an operational principle guiding any modification process associated with the scientific, conceptual framework.

By the mid-20th century the current, conceptual framework for science solidified. Confidence in this framework stimulated a renewed interest in the unification argument. The description of interactions of matter and energy within the microcosm submitted to a 'wave-particle duality' model, a fusion of the particle and wave view for matter and energy. Matter waves became a new way view the propagation of energy; quanta of energy propagating through invisible worlds. As probing within the microcosm progressed, a quantum mechanical theory ascended into acceptance within the scientific community. This theory incorporated experimental results with abstract, intellectual, quantum models. A vision for the atomic and subatomic worlds began to sharpen and draw toward a focus.

The successful predictions for subatomic interactions stimulated cosmologists to incorporate the quantum mechanical theory into their models of description for heavenly events and interactions. New scientific models, proposed compositional and structural patterns for the formation and operation of stars, began emerging. The composition and structure of the Sun became a primary focus for many inquiry models, but the Sun remains a remote and complex entity. There have been successes in modeling the internal structure of the Sun, but most solar models are strongly influenced by conjecture and inference. Modeling drives the inquiry process: it stimulates new questions, and it nurtures a search for believable answers.

By the mid-20th century a segment of the scientific community became interested in an 'evolutionary theory' for the universe. This unification logic included arguments for a sequential, formation of elementary particles, nuclei, atoms, molecules, life forms, planets and moons, stars, and galaxies. However, to call this view of nature a theory is not realistic. There are major information gaps at and within each of these systems. Based on current descriptions and interpretations for operative systems in the universe and an understanding of the scientific, conceptual framework, this unification view of nature would refer to these systems as scientific models.

Scientific evidence supporting either a creation view or an evolution view for the universe is riddled with gaps in scientific information. Therefore, these different views for an origin of the universe beg for a correct use of scientific language. The arguments for both creation and evolution are classic examples for scientific models; neither view qualifies for a status of a scientific theory.

By mid-1950 science education was encouraged to moderate its emphasis of scientific language and vocabulary. The scientific community was calling for an innovative, curricular shift for secondary science. Correlating the classroom activities with the historical development of the conceptual framework was one of the instructional objectives for these new curricula. The positive impact of blending experiential science with conceptual development was a successful, instructional strategy for both science teachers and students. These curricular syntheses use excellent scientific terminology, but vocabulary was not a primary instructional strategy. Whatever the cause, one effect was and is an inconsistent use of scientific terminology by many within the scientific community. Having a scientific theory about 'this or that' scientific effect is surreal. It may be an offering of a scientific hypothesis, but it is not near the mark of being a scientific theory.

The misuse of these terms fosters unnecessary confusion and conflict both within the scientific community and for an interested citizenry. Scientific literacy is a worthy effort, and the scientific community needs to assume leadership in defining its terms, presenting an accurate, consistent use of scientific language, and helping everyone picture the conceptual framework of science.

> Science, Technology, and Simulators

Designing, constructing, and managing the operation of a simulator for an obscure, natural event heightens the level of challenge for scientists. Material scientists must synthesize new materials, inert to variant interactions and foreign conditions. Engineers must design, test, and construct with these new materials. They incorporate new construction techniques with a new generation of instrumentation for detection and measurement. New data acquisition and analysis systems push the inquiry process further into the unknown. New discoveries are carefully sifted for potential use within the marketplace, if not the home. The tangential benefit to the citizenry is a compelling force, impacting every facet of living. Often, perhaps within a decade, scientific advances present the marketplace with new opportunities for application and change.

Technological instruments and experimental techniques present scientists with an opportunity to search events in our universe to unprecedented scales. The use of applied optics and optical systems is central to the probing of the unknown within for most of this work. DNA studies with electron microscopes and other scanning systems provide evidence about defective gene structure associated with debilitating disease. Experimental techniques and procedures associated with this search for identification and resolution are being modified to explore operations within the Sun. The management of digital information by computers is changing the way scientists do science. Though the Sun and a cell are investigative systems vastly different in scale, purpose, and function, computers and optical scanning devises are enabling an extension of scientific inquiry to probe deep within these hidden and previously unknown worlds.

So, science and technology provide reliable platforms for scientists to gather accurate information; credible evidence about interactions and transformations within our natural setting and beyond. Science offer the 'ideal of unification' opportunity for consolidation through its treasury of experimental results. Scientific modeling is moving scientists into virgin fields of inquiry. Conjecture is being replaced with question, hypothesis, and investigation. Experimental evidence is systematically eliminating mystery and intrigue within the universe. Slowly, a shadow masking the unknown is yielding to the light of description and explanation.

Resources for Chapter One: A Prologue On Science

1 Weidner, R. & Sells R., <u>Elementary Modern Physics</u>, (Boston: Allyn and Bacon, Inc., 1960), p 3-9a.

2 Ford, K., <u>Basic Physics</u> , Lexington, Massachusetts: Xerox College Publishing, 1968), p 3-20.

3 Dobbs, B. & Jacob, M., <u>Newton and the Culture of Newtonianism</u>. (Atlantic Highlands, New Jersey: Humanities Press International, Inc., 1995).

4 Koestler, A., <u>The Watershed</u>. (Boston: University Press of America, Inc., 1960).

5 Drake, S., <u>Discoveries and Opinions of Galileo</u>. (Garden City, New York: Doubleday & Company, Inc., 1957).

6 Fischer, R., <u>Science, Man and Society</u>, (Philadelphia: W. B. Saunders Company, 1971).

Others,

Holton, G. & Roller, D., <u>Foundations of Modern Physical Science</u>. (Reading, Massachusetts: Addison-Wesley Publishing Company, Inc., 1958).

Hawking, S. W., <u>A Brief History of Time</u>, (New York: Bantum Books, 1988).

Singer, C., <u>A History of Scientific Ideas</u>, (New York: Dorset Press, 1990).

Chapter Two

Affecters of Science: Scientists, Institutions, and Events

Introduction with a Rumor

'Alexander The Great' was a conqueror of many nations. Among his dreams was the provision of an intellectual bridging to unify ideas and works from the East (Asia) and West (Europe). To this end, Alexander established and provided for the great city of Alexandria, Egypt; his official city, and a pivot between the then East and West. Among the new buildings in Alexandria was a great library. It was to be a residence for the most learned works from both the East and the West. He ordered the transfer of a large number of Greek writings for placement in the library and posted Eratosthenes (the first person to approximate a measurement for the radius and circumference of the Earth) as its chief librarian. Among those documents transferred to Alexandria were the writings from every facet of Greek life; including the writings of Aristarchus, a 2nd century B.C. Greek astronomer.

Years after the death of 'Alexander The Great', the volume of articles submitted for registration and storage at the library began to exceed its capacity and ability to manage. Many of the early Greek writings were on quality parchment and were sold to artists who found them to be an excellent medium for their paintings. In the early 1500's, the economic state for artists demanded a need for an inexpensive medium for their painting. Entrepreneurs, reading the need, found cleaver techniques to remove paint from many older paintings; selling the reclaimed canvass or parchment as a used medium for new paintings.

During the 16th century, the world economy was weak and the outlook for life and living was filled with concern. Most citizens struggled to meet their immediate needs. As often happens in difficult, economic times, artists and musicians in particular were in a dire situation. Musicians had to create new sounds and popular, melodious harmonies. Artists were struggling to find cheap media for their new creations. Periodically, the removal of paint from a canvass or high quality parchment revealed a long-time hidden treasure; occasionally, a historical Greek or Egyptian writing was rediscovered.

Nicholas Copernicus, a monk in the Roman Catholic Church, was interested in the emergence of a new mathematics and astronomy. Rumor purports some of the works of Aristarchus being uncovered during a paint removal process. Copernicus could read and interpret Greek writings; so, the writings were made available to him for assessment and interpretation. Whether this rumor is accurate or not, Copernicus was impressed by the rediscovered

Greek writings of Aristarchus and Aristarchus' view of a Sun centered design for the orbits of the planets. In 1543 Copernicus' heliocentric model for the known Solar System, *On the Revolutions of the Celestial Spheres*, referred to the writings of Aristarchus[1]. The Copernican model was parsimonic, a simplified view for the orderly motion of the known 'heavenly bodies' coursing about the Sun.

The Copernican, heliocentric model was immediately at odds with the geocentric model encouraged by the Roman Church and its intellectual hierarchy. The scientific community was influenced, if not dominated, by teachings from Christian, theological schooling. This bias set the stage for a form of intellectual conflict between differing views for the structure of the Sun, Earth and planets.

As the 17th century emerged, experimental discoveries and new intellectual views for the structural and compositional view of the universe began stimulating change at nearly every juncture of life. Islamic mathematics and astronomy, the Renaissance, and a 'Doctrine of an Infinite Universe' were among the major affecters of science. These are but a few of the major contributions that catapulted scientific inquiry, the arts, and indeed most of mankind into an 'age of enlightenment.' This time became a time of heightened awareness; evoking new questions about nature and launching the scientific revolution.

The intellectual stagnation of the 'dark ages' was reversed by a new view of a mysterious and marvelously ordered universe. Mysticism gave way to pensive inquiry; description, investigation, and interpretation became the hallmarks of scientific inquiry. The formulation and application new mathematical logic, particularly logarithmic logic from the late 16th century and the concept of a limit with the calculus in the late 17th century, stimulated new designs for experimentation and thought. Science applied its developing conceptual framework to enhance emerging ideas like scale, equilibrium, continuity, and boundary conditions and/or limits. These applied, mathematical concepts provided a platform to form definitions for new quantities. Included in a new set of terms to describe an object's physical motion were: time, mass, position, velocity, force, acceleration, and linear momentum.

15th century technology, like the 'blast furnace' technology, propelled the use of iron products into a ready marketplace. The early 17th century agrarian based economy was being infused with new products made from iron. Stronger materials expanded an opportunity to do more, haul more, and contain more. Optical systems: eyeglasses, microscopes, and telescopes clarified the terrestrial view and beyond. These visual enhancements opened new portals of inquiry for composition and structure within the microcosm and the macrocosm.

Many 16th and 17th century scientific investigations were designed by scientists who loved and revered the Lord GOD:

- Copernicus (1473-1543) a 16th century monk/astronomer and a descriptor for the heliocentric model,

- Johannes Kepler[3] (1571-1630), a proponent of the heliocentric model and the descriptor for planetary orbits,

- Galileo Galilee (1564-1642) a proponent of the heliocentric model and the descriptor of trajectory motion near the earth's surface), and

- Isaac Newton (1642-1727) the descriptor and initiator of the theory of mechanics and for models about light and gravitation.

Their inquiry models painted new perceptions for the observed world. Their work stimulated scientists to probe deeper into the unknown. These, and other 17th and 18th-century scientists, forged new procedures and standards for intellectual and experimental inquiry of the natural world.

Heavenly motions were compared with terrestrial motions. Descriptions and understandings of terrestrial motion were extrapolated to offer interpretation for astronomical motions. This connective pathway between the terrestrial and cosmological settings was an initial unifying step for science. It enhanced the value and significance associated with a study and interpretation of events within the terrestrial realm. A thirst for learning about these new manifestations of inquiry and expression became apparent. New scientific discoveries and connections among various scientific events captivated both an interested public and an anxious set of practicing scientists.

The distribution of valued information became a paramount issue; insuring a role for the printing press and the distribution of numerous science books. *Astronomia Nova* and *Mysterium Cosmographicum* by Kepler, *The Starry Messenger* by Galileo, and *The Principia* by Newton are among the primary works galvanizing scientific inquiry and thrusting science on its tenacious course. Information about scientific thought, current discoveries and potential syntheses were more accessible to the scientific community and to a broader profile of the scientifically literate public.

The lives of these early 'affecters of science' reveal common views of God's creation and of the nature of Father GOD. From their book, *Newton and the Culture of Newtonianism*, Dobbs and Jacob make the following comment:

about Isaac Newton: "Newton seems never to have focused solely on the material part of the natural world as modern scientists usually do, but he always remained conscious of the presence of the Deity."[4] Newton and many of his contemporary scientists valued the evidence acquired from searching the physical world as a form of testimony of God's glory. It served to enhance their faith and trust in their Creator GOD. For them, it was a form of circumstantial evidence helping them characterize their GOD. GOD is Who He says He is! GOD did what He said He did! GOD will do what He promised He would do!

This cyclone of synthesis and discovery established the conceptual foundation for all future, scientific inquiries. By the end of the 19th century, the conceptual framework accounting for all physical interactions and transformations was nearly complete. As this century closed the theories of mechanics, thermodynamics, and electromagnetics were formulated and well received by the scientific community. Scientific investigations were yielding results suggestive of a relativistic view for motion. The transition of the three preceding centuries had also produced new conjectures, questions, and hypotheses among scientists. Physical scientists were probing a quantum, structural view for the microcosm and a relativity theory was being formulated.

A Framework for Scientific Inquiry[5]

Following is a brief outline of the foundation for scientific inquiry. Parts of the outline include: 1) a conceptual framework, 2) a procedural framework, 3) a diagram or model depicting an interplay between the content and process components, and 4) a brief comment about applied science and technology.

A Conceptual Framework

Theory >> a scientific statement of commonality about correlated, natural phenomena. Its content and development reflect the axioms of simplicity and generality, present the assumptive and definitive arguments to their highest level of inductive logic. A theory is formulated and presented in an explicit, mathematical form. Important by-products of a theory reside in its ability to summarize experimental results from a multitude of investigations and provide predictive information to form new questions and suggest investigations yet to be designed and performed. While a theory has significant latitude as a predictive instrument, it is also tentative. It is vulnerable to intense testing. While the majority of the scientific community may accept the argument for a theory, other scientists may view this same theory with degrees of skepticism and doubt. The great theories of science account for all known interactions and transformations.

The theories include:
- Mechanics - an explanation of 'how and why' objects move,

- Thermodynamics - an explanation of thermal energy, its affect on matter,

- Electromagnetics - an explanation for interactions and transformations of matter/energy and its response to electric and magnetic effects,

- Relativity - an explanation of the behavior of localized energy propagating at high speeds (greater than 1/7 light speed) and/or its response near a massive body (star),

- Quantum Mechanics - an explanation of the behavior of discrete, localized matter/energy in a resonant environment.

Conservation Law >> a scientific relationship of consistency and/or constancy between defined quantities. The relationship describes interactions and/or transformations within nature. The common, mechanical quantities used to express these calculated relationships include:
- total energy (including energy characteristics: mass and electric charge),

- linear momentum, and

- angular momentum.

Principle / Rule >> an operational statement used to express a consistent pattern, general tendency, or philosophical view typically associated with the state of a system. Examples include the following manifestations:
- equilibrium (action /reaction, equivalence / invariance, symmetry / asymmetry, reversibility / irreversibility),

- complementarity and,

- physical limits (including these principles: Huygens wavelet, Fermat least energy, Heisenberg uncertainty, Pascal pressure, Pauli exclusion, and Bohr correspondence).

Fact >> operational definitions or conventions used to communicate basal information. All of accepted mathematics and science fit within this category. Specific scientific definitions include:
- the fundamental quantities (ex: length, mass, electric charge, and time),

- standards of measurement,

- universal or fundamental constants, and

- a system.

A Procedural Framework Within Science

Model - a mental, mathematical, or physical analog for a natural event.
- **Conjecture** - usually a mental construction, not necessarily based on experimental evidence. Democritus, a fifth century BC Greek philosopher, predicted an atomicity structure for matter. His guess was a beginning for many investigations, searching for information to describe the structure of matter.

- **Hypothesis** - a mental or physical construction formed from consistent evidence; a predictable outcome, based on experimental evidence. From 1803-1807 John Dalton summarized experimental evidence to propose an atomic model to describe the structure of matter.

Modeling and/or Feasibility Study - an inquiry formed by trial-and-error experience. Experimental results are used to check the plausibility of an assumptive logic. This activity is the heartbeat of developmental science; it demands testing and modification until the identification of an acceptable solution. Feasibility studies are a hallmark for product development in enterprising corporations.

Home construction makes use of modeling to market an 'ideal home.' Providing perspective owners with a vision of 'their home' is a powerful marketing strategy. A model home can trigger helpful questions and it presents a visual for possible solutions. This would be a better idea for us. Or, this combination of materials and construction will best accommodate our taste and financial potential.

Modeling, in science, becomes an instigating agent for experimentation and thought. This activity is central to the forces driving scientific inquiry.

A Model for an 'Inquiry / Product' Loop

Some textbooks suggest the 'existence' of a scientific method. This is a very simplistic view of a very complex and dynamic process. There is no single recipe for the systematic investigation of nature. But, the systematic, 'step-by-step,' approach of linear logic has significantly impacted the development of science. Computer programs make use of Boolean operations, 'if-then' logic patterns. This form of investigation is analogous to the steps of scientific inquiry, if … then …logic.

The following illustration displays activity blocks with reversible pathways. There is no prescriptive entry point within this model, but the diagram suggests an importance for a framing of the inquiry question.

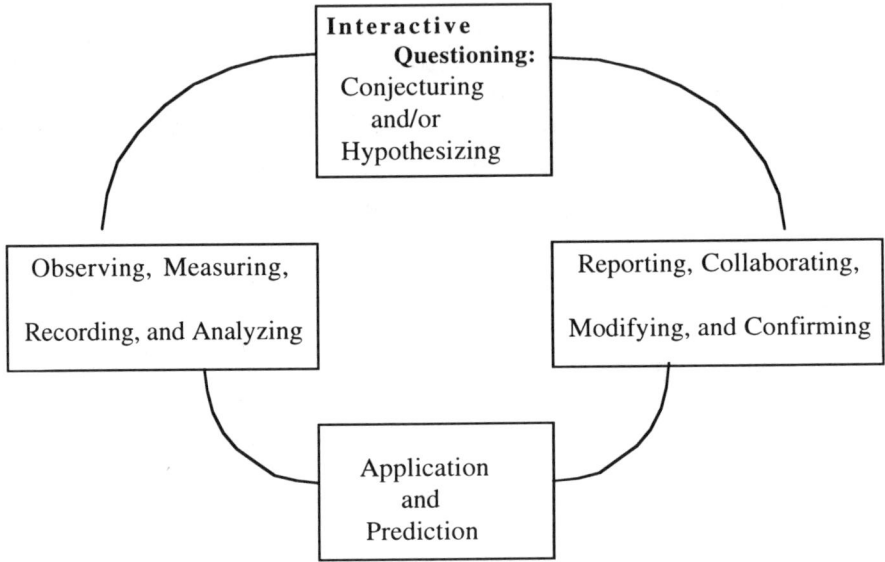

Figure 2.1: An overview of procedure within scientific inquiry.

Unification and the Basic Disciples of Science

The disciplines of science form natural and highly desirable divisions. Specialization within science demands a degree of categorization (systems, investigative techniques, and vocabulary). However, these quisi-partitions can, and often do, form barriers of separation and isolation within science. Partitions tend to obstruct a unified view for science. All disciplines within science have a common agenda: 1) searching nature, replacing the mysterious and the unknown with truthful description and interpretation, and 2) communicating its findings to the scientific community and an interested public. While each discipline of scientific inquiry has its specific challenge, the quest to integrate and incorporate all facets of science into a oneness, a unification of science, remains a goal for many scientists.

Physics assumes a foundational position within the structure of science by virtue of its three fold charge:
- to define and describe the known, physical interactions and transformations,
- to formulate, test, and communicate the great theories of science, and
- to join with the other disciplines of science in the confirmation and application of the scientific, conceptual framework.

All inquiry models depend on the reliability of these frameworks. All evidence and each interpretation is subject to an intense, preview and review process: collaboration, clarification, elaboration and/or modification, and confirmation or rejection.

There are many entry points to contribute to the process of scientific inquiry; yet, there can be only one conceptual platform for the interrogation of nature and the communication of scientific results. The interworkings of science tend to ebb and flow over and through one another. What often seems to be an ending, may really be a new departure for questioning, hypothesizing, and/or investigating nature. Therefore, the development within each discipline contributes to an extension of our understanding and interpretation of events and interactions within the universe. And, there is no most important scientific discipline. Each discipline contributes to a common program; each discipline presents logical arguments for explanation based on observational or experimental evidence.

Occasionally, the structure of science experiences a major conceptual adjustment, an adjustment of limits and/or operational assumptions. This adjustment may clarify an explanation for a complex interaction or transformation observed in nature. The shadow of inference becomes illuminated with each new result; lending support for a trusted interpretation. Each new result brings a cadre of new questions and hypotheses. These questions evoke a need to design new inquiries. As new experimental strategies and/or probing techniques penetrate through the subtle mask of nature, a search for irrefutable evidence confirming another new view of nature begins.

Defining and describing these new and often radical adjustments is very important to science. Historically, these new findings have led science into a new arena of inquiry; a fruitful time of intellectual and experiential inquiry based on question, hypothesis, prediction, and synthesis.

During the 20th century information within each discipline of science has expanded exponentially. Doubling times have been estimated to be within the interval of five to seven years. Computer and optical scanning systems have combined to open new windows for data acquisition and analysis. Complex systems like the human body and the Earth have begun revealing exquisite order; submitting mysticism to the techniques of quantitative analysis.

The scientific tendency to advance unification has become more real and vital. Bioengineering, biochemistry, biophysics, astrophysics, and geophysics are examples of couplings among the scientific disciplines expressing common, operational relationship. The union of physical science with the other disciplines of science signal the importance of mathematical description for/and within natural systems.

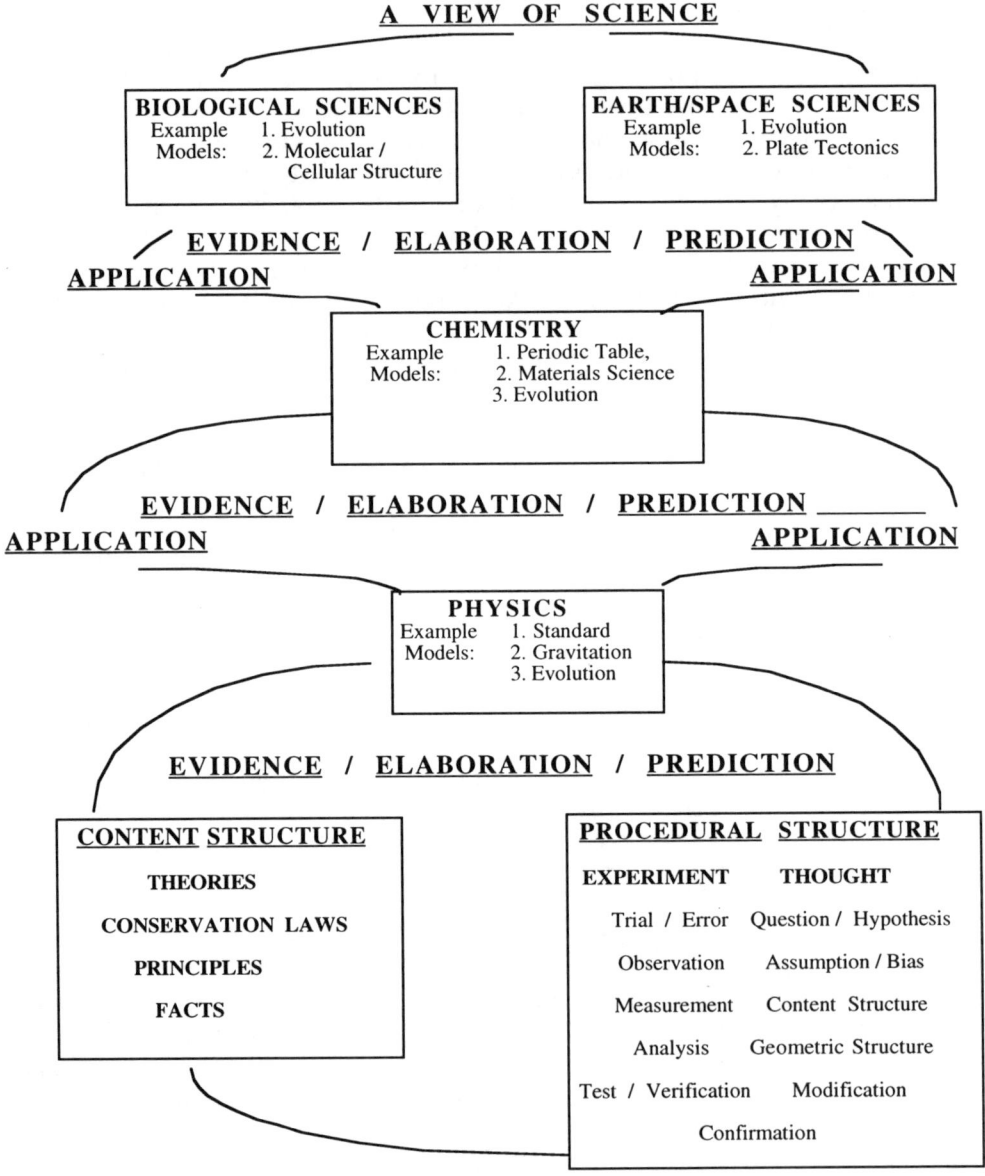

Figure 2.2: A model for the interacting components in science.

A Framework for Applied Science and Technology

Applied science and technology draw their footing from discoveries in the 'pure sciences.' But, they tend to stand quite separate from the pure sciences in purpose. Technology has a primary focus on the design and application of new scientific discoveries. These applications may bring new systems for detection, measurement, data acquisition and/or analysis to both the public and the scientific community. New materials bring new construction techniques and new products. These applications impact the 'quality of life' or issues of human sustenance and tranquillity.

Technology and its multitude of manifestations provide an opportunity to extend our productivity in either of two directions: 1) constructive contributions, those contributions which benefit mankind in a positive way, and 2) destructive contributions, those contributions providing a negative impact on mankind. Today, the medical opportunities afforded a disadvantaged patient are incredibly encouraging. Yet, derivatives of these same medical opportunities can be used by elements of our population in very negative ways; debilitating drugs, biochemical and nuclear warfare devices to name but a few examples.

Applied science and technology embody that aspect of science most directly touching each individual with options of responsibility and choice. Inevitably, optional response among the human population is loaded with opportunity and moral decision. Ethics and morality are not controlled by physical events; they involve a spiritual awareness. And, the personal and corporate responses to these issues and/or events have shaped and will shape the course of mankind's future.

We live in a time of unprecedented opportunity! But, many within the scientific and technological communities would like to separate the 'practitioner of science' from the 'consequences of science.' The dilemma is closely associated with the 'control issue' and that is a very difficult issue for most people. And, so we return an apparent tension between scientific and religious perspectives: Who or what controls the destiny of mankind?

Personal Belief in God and the Practice of Science:

Both religion and science propose a program of personal interaction designed to move the practitioner closer to truth. Religion tends to emphasize the invisible and spiritual truth; while science is absorbed with the determinate and physical truth. As was mentioned earlier, Kepler, Galileo, and Newton were among a group of scientists searching nature to enhance and solidify their belief and personal commitment to their Creator. They viewed their interpretations of natural phenomenon as another level of the beauty and harmony of God's glory and another demonstration of His relationship with the totality of His creation. Their probing of the natural world was in part an attempt to further complete an identification of the visible components associated with the 'glory of GOD.'

The investigation and interpretation of planetary motions in the heavenly realm captivated the early workings of science. These effects are but one of the marvelous, physical events within the realm of God's creation declaring God's glory. By the late seventeenth century, many scientists and theologians began forming a popular variation of this God-centered focus. They began to suggest the natural manifestations of the creation were direct evidence for GOD. Scientific discovery became a proof for theology.

Faith in GOD requires our whole being: heart, soul, mind, and strength.[5] Knowing GOD is to experience GOD; being submissive and obedient to His plan and purpose for His creation. The evidence of GOD is a personal gift and can be acknowledged only at a personal level. In this aspect, science is very different from religion. No one, other than the Creator, can provide the sustaining evidence of His existence and His desire to be acknowledged personally. Should a person or persons choose to share their personal experiences with others (witnessing to and/or evangelizing with others), they can encourage their friend, but forming a comprehensive and meaningful relationship with GOD is a unique, personal experience for each of us.

The members of the Church, the community of believers faithful to Christ Jesus, serve and encourage one-another in a ministry of reconciliation with GOD and with our neighbor. This manifestation of God's grace, "We love because He first loved us."[6], is reflective of a life focused on: praise and worship to a Redeeming King, a belief and study of God's Holy Word, and a witness of life experience. A Christian acknowledges a life perforated with sin, a need for reconciliation with GOD through confession of sin, and an acceptance of His forgiveness. A believer is a willing partner in the Sacrament of the 'Body and Blood' of our Savior, Christ Jesus; a sinner being redeemed and standing in the hope of a transformation into the 'image of GOD' at one's eternal salvation. An acknowledgement and acceptance of GOD is the most important personal decision one will encounter in this life. Is that a statement of fact, or is that an example of evidence? For some, it is neither?

Science follows a course of searching out evidence to describe the physical world and providing consistent interpretation for the manifestations of God's glory. Both committed believers and equally adamant unbelievers have provided significant contributions to the development of science. Both personal and corporate effort advance explanation and understanding of the operation and orchestration of events within the universe. Yet, all scientific evidence and results must be shared, verified, and accepted within the community of science. Science may, and often is be done by individuals; but, the incorporation and acceptance of science is a community experience. And, its incorporation does not guarantee its truthfulness. No individual scientist or group of scientists 'hears and/or sees' all there is to observe and describe within nature.

Today, many people identify with a quote from the front, cover page of Hugh Ross' *The Fingerprint of God*: "Recent scientific discoveries reveal the unmistakable identity of the Creator."[7] Ross' argument for a Creator GOD is compelling and refreshing. This work reviews the scientific description for the created world and presents a conciliatory argument to align the 'six days' of the creation given in the 'Word of GOD' with the six, major intervals of change expressed by science through geologic time. This view facilitates a scientific view for creation and a focus on the Creator.

Chapter Two Resources
Affecters of Science: Scientists, Institutions, and Events
  1. Durant, W., The Story of Civilization: Part II, The Life of Greece , (New York: Simon and Schuster, Inc., 1939), p 669.

  2. Drake, S. (1957). Discoveries and Opinions of Galileo, (Garden City, New York: Doubleday & Company, Inc., 1957).

  3. Koestler, A., The Watershed, (Boston: University Press of America, Inc., 1960).

  4. Dobbs, B. & Jacob, M., Newton and the Culture of Newtonianism, (Atlantic Highlands, New Jersey: Humanities Press International, Inc., 1995)

  5. Barker, K. (General Editor), New International Version, (Grand Rapids, Michigan: Zondervan Bible Publishers, 1985), Mark 12:29-31.

  6. ibid. I John 4:19

  7. Ross, H., The Fingerprint of God, (Orange, California: Promise Publishing Company, 1991).

Chapter Three
Mathematical Modeling: A Process of Analysis and Synthesis

Introduction:

The good fortune and/or skill of framing an evoking question about **nature** is central to scientific inquiry. Questions about a given phenomenon demand the exacting of a response. Hypothesizing about the question, isolating and observing the phenomenon, gathering data, analyzing the data, rendering an interpretation, and communicating a result are significant components associated with an inquiry process. Each of these vital steps is intense; they are riveted with questions, driven by interrogation and testing. Each step within the scientific inquiry adds to a crescendo of information about a specific event. The end for a scientific inquiry is reflected in a communication of its reliable result.

A mathematical model is a culminary expression for an experimental result. A mathematical model is a predictive platform to extrapolate a result beyond the boundaries of the investigation. A mathematical model is a trusted platform of succinct communication. Scientific literacy demands its use. It is a vital element of communication among scientists, engineers, mathematicians, and a scientifically literate public. The following terms are axiomatic to a scientific inquiry:

- **parsimony** - a frugal, simple expression for an ordered set of observations and/or results. The framework for this expression includes: 1) an objective or question for the inquiry, 2) an outline of assumptions to gather descriptive information, and 3) an analysis of numerical data with a result expressed in a mathematical model.

- **induction** - an accumulative description (parts contributing to form a comprehensive whole) given for an event.

- **deduction** - an interpretation given by identifying, isolating, and describing the contribution of each component (a whole view reduced to an investigation of interrelated parts) to the operation of a system.

Again, there is no best way to do science. But, the analytical techniques described in this chapter have stimulated an asking of questions and have successfully guided many scientific inquires. This chapter has three parts:

1) Style and Technique in Mathematical Modeling,

2) A Historical Example - The Kinetic Molecular Model, and

3) A Closer Look at Universal Constants.

Style and Technique in Mathematical Modeling

Mathematical modeling is a process of forming a mathematical description among the variables used to define the state of a system. The acquisition of experimental data, its tabular display, and a judicious comparison of variables initiate the analysis. Most simple graphs paint a comparison for two variables, a dependent and an independent variable. If a system has more than two variables, then changes in the state of the system are systematically controlled by: 1) defining a fixed or constant value for each, additional parameter, and 2) collecting data describing the variation for the two quantities. This comparison process continues until all of the variables can be separated and mathematically ordered into two separate functions. Mathematical order and operation are modified until a relationship reflects the operation of the system; and, the mathematical result forms into a linear relationship.

A linear relation is an expression of the operational variables within the system and assumes a slope-intercept-form for a straight line, $y = mx + b$.

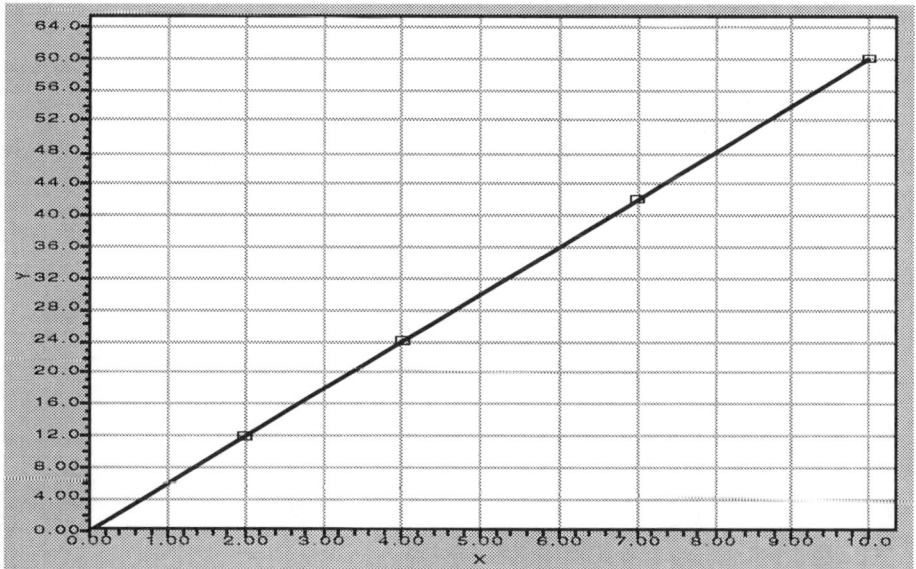

Graph 3.1: An example linear variation with variables y and x.

The mathematical expression for this relationship is called a direct variation. The values for the variable y are said to be dependent on values for the variable x. The format for this dependency is expressed as follows: 1) $y = f(x)$, 2) $y \propto x$, and 3) $y = mx + b$. The independent variable, x is the standard of comparison in an inquiry. The independent variable controls the structure of the data table and the formation of the graph. Also, it may be linked to an earlier investigation, or it may reflect the observer's confidence in the accuracy of measurement among the system variables. The slope of the straight line signal the 'rate of change' between the two variables, $m = \Delta y / \Delta x$. Or, this comparison defines the 'rate of change' among the mathematical combination of those variables assigned to form a dependent function and an independent function.

The constant **b** is the intercept value for the ordinate (dependent function) axis when the independent variable and/or function is zero. If the dependent (ordinate) function and the independent (abscissa) function both approach the-limit-of-zero, the argument is said to be a direct variation or, a direct proportion. For this argument the intercept is zero and the general form of the equation becomes: **y = m x** with b = 0, the y-intercept when x=0; and, the mathematical model for **Graph 3.1** is, **y = 6 x.**

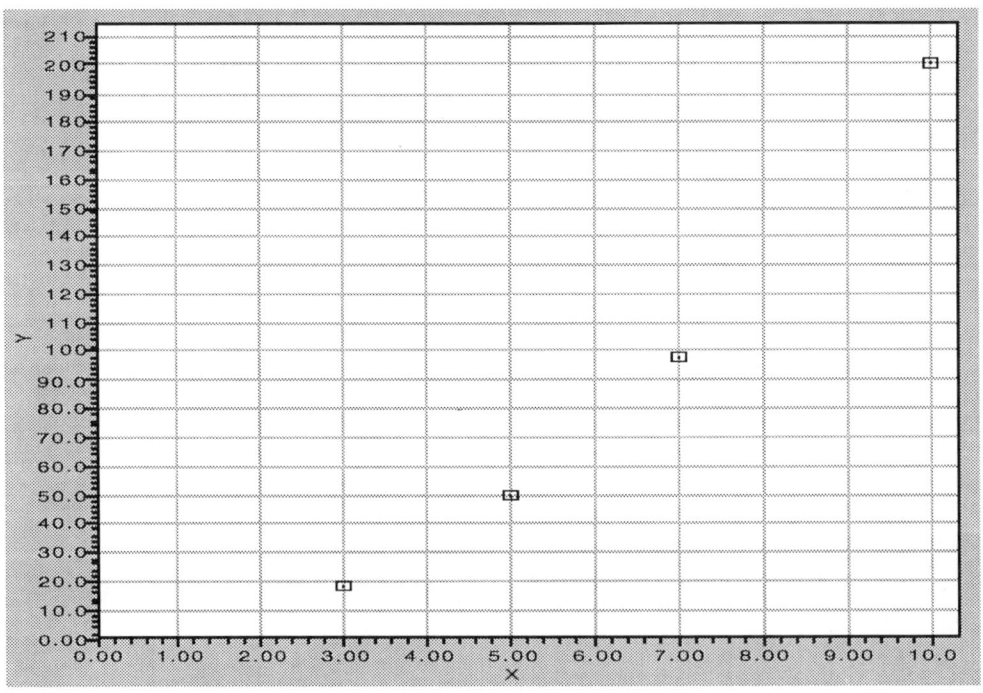

Graph 3.2: A non-linear, second order function of variables **y** and **x**.

Often a graphical comparison of empirical data does not form a straight line. To linearize the comparison, mathematical operations must be performed to the variable(s) associated with the dependent function and/or the independent function. **Graph 3.2** is a representation for an example set of data. This example was designed to form a parabola, a second order function. By comparing mathematical functions for both the ordinate and abscissa axes, a linear variation can be formulated. Understanding this linearization process requires mathematical skill and experience by both the analyst and the reviewer. But, this succinct platform of communication is an expected norm for sharing an experimental result within the scientific community.

To linearize the result expressed in **Graph 3.2**, a modification of the independent variable to x^2 **-values** is recommended. Then a comparison of the original **y-values** and the x^2 **-values** can be checked for linearity; $y = f(x)$ and $f(x) = x^2$.

x	x^2	y
3	9	18
5	25	50
7	49	98
10	100	200

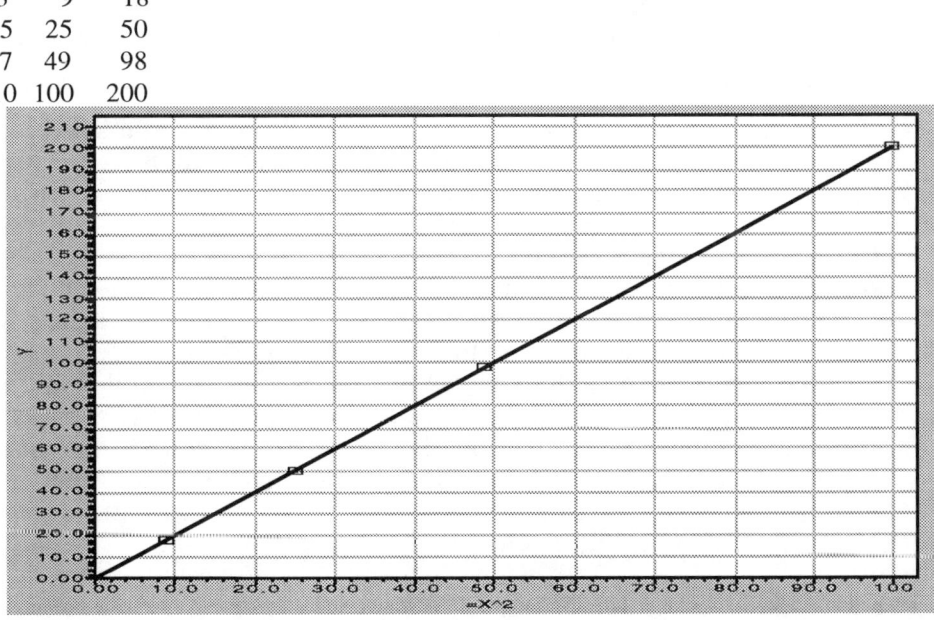

Graph 3.3: A linear variation of **y** and **x** formed by a modification of the variable **x**; $f(x) = x^2$.

This graphical comparison, $f(y) = y$ and $f(x) = x^2$, produces a linear variation with the 'rate-of-change' of two, $m = \Delta y / \Delta x^2 = 2$. The mathematical model becomes the equation for the line, $y = 2 x^2$ and also, the equation for the second order function displayed in **Graph 3.2**, a parabola.

Considerations for a Graphical Analysis:

- **Does the graph appear to be linear?**

    1) If **YES**, then form a 'best-fit' straight line through the points and
       determine the mathematical form for the variation.
       a. Does the distribution include the origin?
           1. If yes, then the variation is a direct variation.
           2. If no, then evaluate the 'y-intercept'; determine its meaning
              or association.
       b. Evaluate the 'rate of change' for the linear variation,
          $m = \Delta [f(y)] / \Delta [f(x)]$.
       c. Define the 'equation-of-the-line' according to the general form for
          a linear variation:
           1. a direct variation, $y = m\,x$; or,
           2. $y = m\,x + b$.

    2) If **NO**, then modify one (usually the function including the indepen-
       dent variable) or both functions until a linear variation is identified.

- **What is the significance of the 'rate of change' in a linear variation?**

    Understanding the 'rate of change' for the system variables typically
    has played an important role in providing an interpretation for an experi-
    mental result. Any experimental values held constant during the
    investigation must be mathematically associated with this value. Con-
    stants may be either 'explicit' or 'implicit'. The definition of explicit
    constants, experimental parameters, provides an opportunity to control
    change within a system and collect orderly information for each, system
    variable. Implicit constants are often associated with the experimental en-
    vironment. To recognize their contribution to an experimental result, each
    parameter must be identified, isolated, and evaluated. These masked
    quantities are linked together by the grouping principles (associative
    and/or commutative) of multiplication. This idea is among the more
    subtle and powerful ideas in science. And, it is a pathway to new views
    and/or discoveries within nature.

- **When is an investigation or inquiry complete?**

    When a believable continuity between the experimental question, hy-
    pothesis, collection and analysis of data is established, the interpretation
    can be formulated. The rate-of-change, including operational units, in the
    mathematical model becomes the focal point for identification and expla-
    nation. A deductive process separates the known constants from the ex-
    perimental constant, attempting to explain or account for the remaining
    factor.

If multiple inquires provide evidence of a common, indivisible factor, this quantity is isolated and defined to be a unique constant, an **universal constant**. A universal constant is to science, what a prime number is to mathematics.

A Historic Example: A Kinetic, Molecular Model For A Gas[2]

An excellent example of mathematical modeling was given with the formation of the 'ideal gas equation'. This model offered a partial description for the behavior of a contained gas using the variables: pressure (**P**), volume (**V**), moles of molecules (**n**), and absolute temperature (**T**). The description for this closed system used deductive logic; isolating the system response into three, distinctive operations: 1) Boyle's Principle, 2) Charles' Principle, and 3) Gay-Lussac's Principle. The interrelationship between the experimental variables and parameters from these three inquiries forged a comprehensive, mathematical interpretation describing the behavior of a contained gas. James Maxwell and Ludwig Boltzmann united the 'idea gas equation' with Newtonian and statistical mechanics to step toward a synthesis of the 'theory of the thermodynamics.'

>> Boyle's Principle

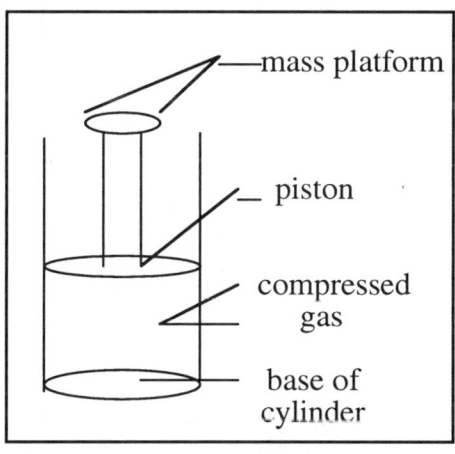

Figure 3.1:
A representation of the apparatus one might use to investigate Boyle's Principle.

In 1662 Robert Boyle completed a first step in this experimental trilogy by summarizing his investigation with the relationship, $PV = k$. The investigation was performed at nearly constant temperature, room temperature; and, its connection with the experimental constant 'k' remained 'implicit' until the early 1800's. Using an apparatus similar to the representation in Figure 3.1, Boyle used a closed system to collect data describing the relationship between variations in gas pressure and gas volume.

As mass was added to the platform its gravitational force caused the volume of gas to diminish. The reduction in gas volume generates a simultaneous increase in gas pressure. If the process of compression is done slowly, the temperature of the gas remains nearly constant. Today, this process is commonly called an adiabatic compression of an 'ideal gas.' As the pressure of the confined gas increases, the volume of the confined gas decreases. This form of variation signals an inverse variation between the gas pressure and volume.

Boyle's laboratory investigation was done under normal, experimental conditions. One implicit constant associated with the inquiry process was the temperature of the confined gas; its temperature remaining nearly constant at room temperature. The following data and graph are analogous to his experimental result for a variation of the gas pressure for a confined gas as a function of its changing volume.

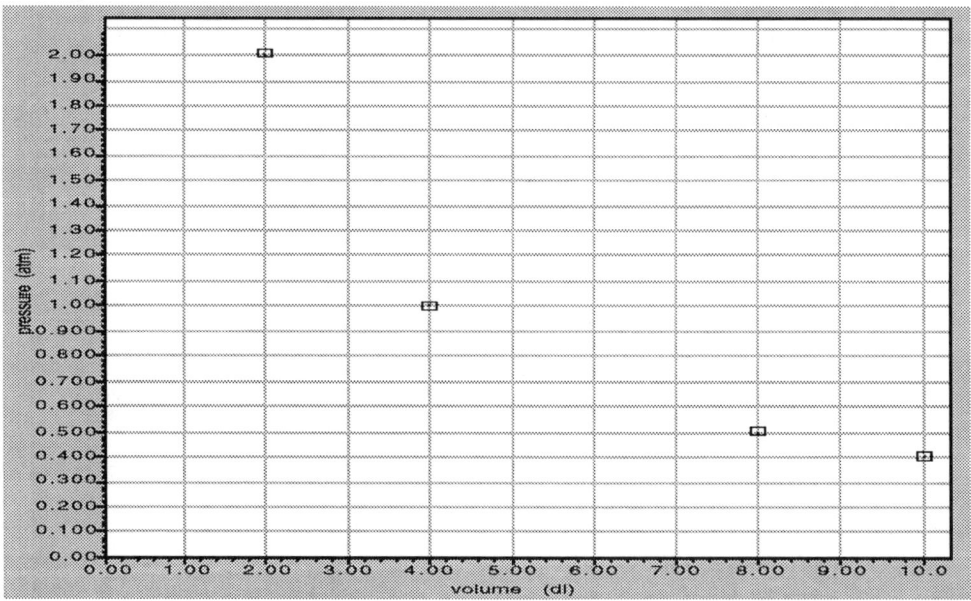

Graph 3.4: Pressure as a function of volume for a confined gas at constant temperature.

Graph 3.4 has the characteristic shape for an inverse variation. To verify this inference, a modification to the data must be exacted. To linearize an inverse variation, the usual mathematical operation includes performing the inverse operation on the independent variable and comparing the gas pressure with the modified argument, the inverse of the gas volume. This modification and comparison is presented in **Graph 3.5.**

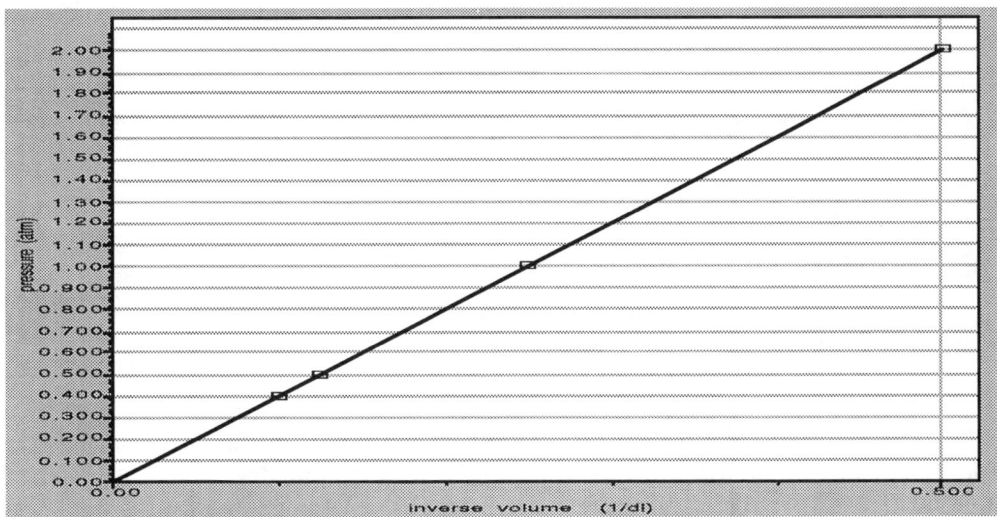

Graph 3.5: Comparison of confined gas pressure and the inverse of volume.

[**NOTE:** A mathematical modification is usually exacted on the independent variable, a quantity often bearing a higher confidence in accurate measurement. This operational norm is not rigid in all applications; rather, it remains a guideline for mathematical modifications administered to the data and forming a mathematical model.]

The analysis of the 'P-V argument' proceeds with the following mathematical notation:
- gas pressure is proportional to inverse gas volume, and
- $P = k_1 [1 / V]$.

What is the significance of the experimental constant, k_1?

  a. the slope (the 'rate-of-change' among the variables) for the straight line:
  $$k_1 = \Delta P / \Delta (1/Vol),$$

  b. an analysis of units:
  $$k_1 = \text{atmosphere - deciliter (atm-dl), with}$$
  $$k_1 = \text{work done (area under 'P-V' curve), and}$$

  c. by 1810, Boyle's implicit temperature was modified; making the temperature an explicit variable:

  $$k_1 = k_2\ f(T)$$

k_1 --> explicit, 'temperature', and

k_2 --> implicit, experimental parameters.

The constant of variation (proportionality constant) must include a functional expression for temperature and energy, work done on the closed system. The following argument is one, logical expression for these quantities:

P - pressure (compressed gas)
V - volume (compressed gas)
F - force (piston)   • $P V = (F/A) (A h) = F h = W$ and
T - temperature (compressed gas)
A - cross-sectional area (cylinder)   • $k_1 = k_2 f(T)$
h - height of piston plate from base
W - work done by compressed gas

To say the temperature was constant during the investigation done by Boyle is to make a significant assumption. It is probable that the temperature was nearly constant (room temperature); but, Boyle's interpretation makes no mention of the significance of constant temperature. Therefore, this constant was actually an implicit constant, an experimental parameter having a nearly constant value and under the control of experimental conditions.

More than a century was to pass before Jacques Charles and Joseph Gay-Lussac independently modified Boyle's apparatus to study the effect of temperature variance on the behavior of a confined gas. Considerations, hypothetical arguments, about hidden interactions within the structure of matter paved the way for the investigative strategy of Charles' and Gay-Lussac's inquiries. Among those contributing to this effort[2] were the following scientists:

• Antoine Lavoisier (1789) - experimental work with calcination and the beginnings of charting chemical reactions for oxidation and reduction,

• J. L. Proust (1797) - identification of the principle of 'definite proportions' for a given compound, and

• John Dalton (1800+) - delineation of the principle of 'multiple proportions' for the combinations of the same elements to form different compounds and a proposal of an atomic model for the structure of matter.

>> Charles' and Gay-Lussac's Principle

During the last decade of the 18th century French scientists and hot-air balloon enthusiasts, Jacques Charles and Joseph Gay-Lussac independently extended the work of Boyle; studying of the effect of the volume of gas on the temperature at constant pressure. The constant pressure was the composite of the following components: 1) atmospheric pressure, 2) pressure due to the liquid plug's interaction with gravity, 3) the vapor pressure from the liquid plug, and 4) the adhesive force between the plug and the glass tubing. They used an apparatus similar to the one presented in the following diagram.

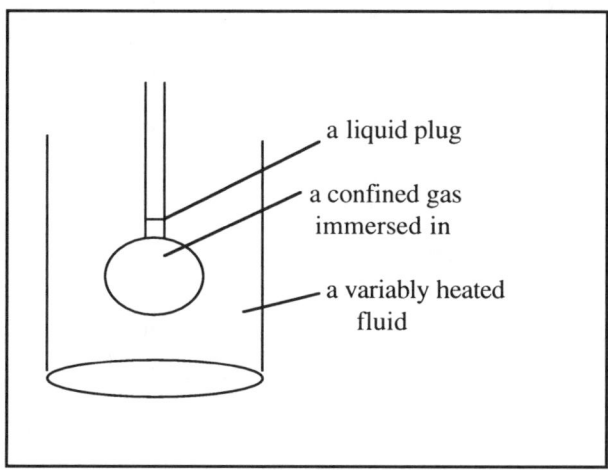

Figure 3.2: Charles' modification to the Boyle investigation, V = f(temp) with an implicit, constant pressure.

Like Boyle's experiment, Charles' investigation was performed with an implicit, experimental constant, pressure. Also, because the gas was confined during the experiment, the number of molecules forming the investigative sample was also an implicit, experimental constant. In about 1808 Gay-Lussac formalized this insight by recognizing the definite combining volumes of hydrogen and oxygen to form water vapor.

In 1811 Amedeo Avogadro extended Gay-Lussac's interpretation for water vapor to establish his definition of a mole to formulate Avogadro's constant, a hallmark among scientific inquiry. With Avogadro's stunning result, circumstantial evidence for Dalton's earlier hypothesis about the atomic and molecular structure of matter was enhanced. Within the span of two decades (the decade closing the 18th century and the decade introducing the 19th century), a summation of intense, experimental investigation and shared scientific results began lifting one of nature's secrets, the scale and structure of the microcosm.

The proposal of a 'kinetic-molecular model' galvanized science clearly demonstated the utility of mathematical modeling as a tool of scientific analysis and synthesis. The guidelines for mathematical modeling were advanced by the work of Charles and Gay-Lussac Principle. Following is a graphical analog summarizing this principle:

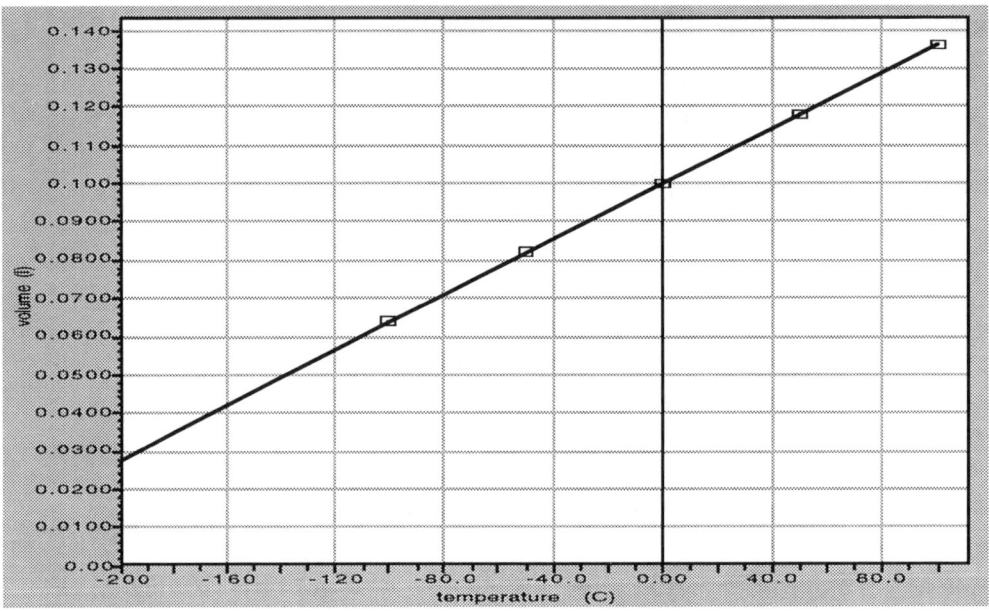

Graph 3.6: The expansion of a confined gas under nearly constant pressure.

This is a linear variation with the change in volume being proportional to the change in temperature at a constant pressure. This is the basic idea assoc- iated with the operation of most thermometers. Neither Charles nor Gay-Lussac recognized the mathematical connection between the three variables: volume, temperature and pressure. But, Gay-Lussac was able to press the logic forward to open a particle view for the world of atomicity. This view was readily grasp by Avogadro; yielding his definition for a mole of matter. Another application proposed by Lord Kelvin established the 'absolute temperature' scale.

The description for change within a gaseous, closed system shifted from a qualitative view to a quantitative view with an incorporation of the kinetic ener- gy for molecules being proportional to the absolute temperature. One, common mathematical model used to describe a state for an ideal, confined gas follows:

$$P V = R n T, \text{where}$$

R - ideal gas constant, R = 8.31 J/mole-° K and
n - the number of moles (one mole equals 6.023E23 things).

By the middle of the 19th century a culmination of experimental results was preparing the platform for another of science's great syntheses. Hypotheses and predictions were being confirmed. Some of the major contributions leading to the formation of the theory of thermodynamics include:

- Sadi Carnot (1824) - evaluation of work done during a reversible cycle,
- James Joule (1840's) - quantification algorithm for the transfer of thermal energy in mechanical systems,
- Julius Mayer (1842) - hypothesis for the conservation of all forms of energy,
- William Thomson (1848) - proposed the Kelvin temperature scale,
- J. W. Gibbs and Ludwig Boltzmann - used statistics to describe atomic and molecular events,
- James Maxwell (1858) - proposed an argument for gaseous states having 'degrees of freedom' associated with a partition of energy.

These investigations led scientists to a model describing the behavior of matter in a gaseous state and under a broad set of conditions. Inquiry models investigating the other physical states of matter (liquids and solids) confronted more restrictive states. The mathematical models used to describe limited change were significantly more complex then the solutions for an 'ideal gas.' But, special points, like the triple point for water, began yielding evidence of a relationship between composition and structure for these states of matter. In particular, the experimental work done by Joule (transformation of mechanical work into thermal energy) and the studies about 'specific heats' began painting connections for the operative principles of a new theory; a theory for thermodynamics. These principles included:

- energy conservation during all interactions and transformations,

- changes in the internal energy within a system tending to alter its entropy state at a given absolute temperature,

- energy transformations are irreversible with thermal energy being dissipated at each interface.

A Closer Look at Universal Constants

>> The Classic Electron Radius and Coulomb's Experimental Constant

In 1785 Charles Coulomb designed an apparatus to investigate electric force, electric charge, and charge separation. Coulomb's mathematical model gave an interpretation for isolated electric charge. The classic form for this important force and its components is given as follows:

$$F_e = k \, q_1 \, q_2 \, / \, r^2$$

The experimental constant, $k = 8.998E9$ (kg/C^2) (m^3/s^2), may encapsulate combinations of both explicit and implicit values. The structure of the unit associated with this value is very important. Reducing the unit to its fundamental components and regrouping those components may signal various options for consideration and calculational trial. From the above unit components two prominent ideas are discernible: 1) a mass-charge ratio, (kg/C) $(1/C)$ and, 2) a connection with a mass element in an orbit, (m^3/s^2).

An experimental constant must either be a universal constant or a collection of other constants. At the time Coulomb conducted his initial inquiry, he did not have the information base of modern physics. A quantum view of matter, charge, space-time, and energy has provided a platform to predict many particle characteristics. Some of these particles have been neither observed nor measured. Isolating information associated with a constant's unit can stimulate the formation of questions, perhaps questions for unification:

  • the quantity (m^3 / s^2) could include a length and a speed squared,

  • an electric charge is quantized, $q = n\,e$, where 'n' is an integer and 'e' is the elementary, electric charge, $e = 1.602E-19$ Coulomb,

  • electron mass, $m_e = 9.11E-31$ kg, and the Coulomb, 'C', being the unit of measure for electric charge,

  • the speed of propagation for electromagnetic, radiant energy, is the speed of light, $c = 2.998E8 m/s$, and

  • a new symbolic representation for this unit might include:

$$(m_e / e^2) (c^2\ l) = 8.988E9\ (kg/C^2)\ (m^3 / s^2).$$

The unidentified quantity in this expression is the parameter 'l', a length of probable significance. Again, isolating this quantity yields:

$$l = 8.988E9\ (kg/C^2)\ (m^3 / s^2) / [(m_e / e^2)\ (c^2)],$$

$l =[8.988E9\ (kg/C^2)(m^3 / s^2)][(1.6E-19C)^2]/[(9.11E-31\ kg)(2.998E8\ m/s)^2]$, and $l = 2.817E-15$ m, a value often associated with the classical, electron radius.

Another connection worthy of noting is an electron orbital and its associative characteristics. Kepler suggested a relationship for an orbital separation and an orbital period, r^3 / T^2. If one divides the quantity '$c^2\ l$' by the factor $4\pi^2$, another orbital characteristic is identified:

$$k_{atom} = r^3 / T^2 = [c^2\ l] / [4\pi^2] = [251.058\ m^3 / s^2] / 4\pi^2 = 6.359\ m^3 / s^2$$

This value is a Keplerian analog for the orbital values associated with the Bohr model and its description for elemental hydrogen.

A Unification: The Planck & Universal Gravitation Constants

Max Planck and most of the masters of physics clearly understood the powerful framework of mathematical modeling as he synthesized his famous constants. Time and intense inquiry will illuminate the true significance for these values. One of the modern models being tested and modified is called the 'standard model.' It makes use of these parameters to evoke some very interesting questions and generate interesting hypotheses for testing and investigation. Again, it makes use of this concept of unmasking an implicit constant. Following is one possible synthesis for these factors:

$G = 6.672\text{E-}11 \ (m^3/s^2) \ / \ kg$ and $h = 6.626\text{E-}34 \ kg \ m^2/s$

- $G \ / \ h \quad = [6.672\text{E-}11 \ (m^3/s^2) \ / \ kg] \ / \ [6.626\text{E-}34 \ kg \ m^2/s]$,

$\qquad = 1.00709\text{E}23 \ (m/s) \ / \ kg^2$

$c \ / \ m^2 \ = \ 1.00709\text{E}23 \ (m/s) \ / \ kg^2$,

$m^2 \ = \ 2.998\text{E}8 \ m/s \ / \ 1.00709\text{E}23 \ (m/s) \ / \ kg^2 \ = 2.977\text{E-}15 \ kg^2$

$m \ = \ 5.456\text{E-}8 \ kg$, a calculated, unification mass[3]
$\qquad\qquad$ in the GUT model[4].

- $G \ h \ = [6.672\text{E-}11 \ (m^3 \ / \ s^2) \ / \ kg] \ [6.626\text{E-}34 \ kg \ m^2 \ / \ s]$,

$\qquad = 4.4202\text{E-}44 \ m^5 \ / \ s^3$

1) $c^3 \ l^2 = \ 4.4202\text{E-}44 \ m^5 \ / \ s^3$

$\qquad l^2 \ = \ [4.4202\text{E-}44 \ m^5 \ / \ s^3] \ / \ [2.998\text{E}8 \ m/s]^3 = \ 1.6404\text{E-}69 \ m^2$

$\qquad\quad l \ = \ 4.0502\text{E-}35 \ m$, a length referred to as the Planck length and,

2) $c^5 \ t^2 = \ 4.4202\text{E-}44 \ m^5 \ / \ s^3$

$\qquad t^2 \quad = \ [4.4202\text{E-}44 \ m^5 \ / \ s^3] \ / \ [2.998\text{E}8 \ m/s]^5 = \ 1.825\text{E-}86 \ sec^2$

$\qquad t \quad = 1.351\text{E-}43 \ sec$, a time referred to as the Planck time and hypothesized to be associated with a unification of all known forces[3].

This chapter summarizes a method of questioning nature, a methodology to identify and define new quantities and suggest new avenues of inquiry. Mathematical modeling has been used historically to describe the interrelationships among system variables and to hypothesize about previously unconsidered questions, connections, and possible inquiries. It can provide a pathway to connect both earlier and new, correlative investigations. These pathways have and will form bridges for unification and simplification in science. Mathematical modeling is a powerful tool of expression as science advances its quest for truth and an understanding of nature's magnificent order and subtle mysteries. This method of modeling offers its practitioner an opportunity to probe deeper into the unknown. It provides direction to systematically remove the sophisticated masks within nature; illuminating the hidden harmony encompassing nature's composition, structure, and exquisitely created purpose.

Part II of this book introduces an argument for a quantum view of gravitational events within the Solar System. Orbiting satellite systems provide a grand geometric setting to frame questions and form hypotheses accounting for the beauty and perfection of this created order. Harmonic resonance, destructive interference, and an inferential model guide a computer search for linkage with astronomical data describing the characteristics for an orbital reality. The search makes significant use of decoding fundamental constants among gravitational events within the Solar System.

Resources For Chapter 3:

1. Ford, K.W., <u>Basic Physics</u>, (Lexington, Massachusetts: Xerox College Publishing, 1968), p 3-20.

2. Holton, G. & Roller, D., <u>Foundations of Modern Physical Science</u>, (Reading, Massachusetts: Addison-Westley Publishing Company, Inc., 1958), p 364-459.

3. Misner, C., Thorne K., & Wheeler, J., <u>Gravitation</u>, (San Francisco: W.H. Freeman, 1973), p 428.

4. Lederman, L., & Schramm D., <u>From Quarks To The Cosmos</u>, (New York: Scientific American Library, 1995), p 172.

Other:

Beatty, Metzdorf, Reid, Shepherd, Yamaguchi, Sundaram, <u>Interactive Methods for Selected Topics in Physics & Mathematics Using the Computer</u>, (Boulder, Colorado: The University of Colorado, 1973).

[**note:** The author was the principal programmer for the mathematical modeling program >> **MODEQ** in this **NSF** project.]

Part II: An Example of Science

Cosmologists search for scientific evidence among the heavenly events and processes. They design models, offering connections between the vastness of the macrocosm and planet Earth. A correct 'seeing' of the interactions in deep space present opportunities to hypothesize about the structure and composition of the macrocosm and the universe. Among the most elusive quandaries within this realm is an understanding of the nature of gravitation. Its effect is associated with all masses. The scale of interaction with large masses in the macrocosm is an awesome event to witness and a marvelous event to experience.

Quantum gravitational effects are likely associated with double stars, near neutron stars, and/or in the environment near black holes. The clues for a quantum, gravitational effect are subtle and these interactions are remote and difficult to study. However, scientific confidence in the existence of gravitational waves remains high. New models will likely clarify the nature of gravitational force, and provide insight for a unification of nature's four forces: the strong and weak nuclear forces, the electromagnetic force, and the gravitational force.

The satellite motions within the Solar System, orbital elements in synchronous motion about a primary mass, reflect a form of stability and present an experimental environment capable of scientific interrogation. If gravitational waves exist, then structure and composition within the Solar System should signal the effect of weak quantum, gravitational interactions. Subtle evidence for wave characteristics should exist, but require a different view for structure within this awesome system. 'Seeing' the Solar System from a new, descriptive perspective includes viewing the Solar System as:

- a resonant, gravitational system,

- a quantum view for mass, space, angular momentum, and total energy, and

- harmonious couplings between: 1) the Sun and the planets, and 2) a planet and its moons.

Scientific publications are filled with examples of good science. The chapters in Part II of this book will pursue a strategy for the development of a quantum, gravitational model for the structure and dynamics of orbital elements within the Solar System. Following is a brief summary for four of the five chapters (chapter eight is a closing comment) in Part II:

Chapter Four - An Example of Science in Progress

Resonant, gravitational effects within the Solar System are numerous. One of those effects is the dynamic motion associated with the Solar System's barycenter. The periodicity for this event may clock the interactions for the complete, solar environment. The center of mass for the Solar System is known to reside predominately within the limits of the Sun's photosphere. Periodically the barycenter may move outside the radial limit of the photosphere; and, this iterative motion may affect the internal operation of the Sun, promoting ancillary effects on each of the planets.

Chapter Five: Stepping Toward a Resonant, Gravitational Model in the Solar System

A structural connection between the Coulomb and the Cavendish constants is explored. Analogous components within these experimental constants and the logic of the Bohr model motivate the formulation of a wave, interference model for the planetary orbits.

Chapter Six: A Wave, Interference Model for Gravitation in the Solar System

The ellipse is an extraordinary example of perfect symmetry. Yet, Kepler's first principle for a planetary orbit is an asymmetric statement. A geometric modification for this asymmetry provides an entry step to form a gravitational wave interference model for planetary orbits within the Solar System. Could motion within the Solar System be governed by quantum, gravitational waves?

Chapter Seven: A Quantum, Gravitational Model for Satellite Motion in the Solar System

A computer search for quantum effects within the Solar System analyzes the following orbital characteristics:

- foci separation, $2c/n$,

- angular momentum, twice the orbital area per unit time ($2A/T/n$),

- incremental mass within an orbital system, m_0/n,

- speed, $n\,v_p$

- period, T_p/n^3.

Within this chapter a proposed partition of the Cavendish constant, $G = 6.672E\text{-}11 Nm^2/kg^2$, would make this, gravitation constant an experimental constant.

Chapter Four:
An Example of Science In Process

Overview:

Progress in science is often a systematic, 'step by step' process. As an interrogation of nature culminates with an interpretation, summaries are published and the process of review and verification begins. A principal hope associated with any inquiry process pivots on a favorable review and acceptance from the scientific community. This process of acceptance and confirmation by scientists serves to validate prior investigations and project toward future inquiries. Confirmation may provide an important platform to frame new questions, form new hypotheses, and design new inquiry models to extend the probing process. The following argument serves as an example of this process. Two interests stimulated the design of this inquiry during the summer of 1993:

- a search for astrophysical phenomena demonstrating resonant characteristics within the Solar System, and

- investigating the dynamics of the 'center of mass' of the Solar System with respect to the center of the Sun.

Earlier computer models[1] describe the motion of the barycenter for the Solar System and infer a dependency on an orbital harmony of the planets. The motion of the Solar System's barycenter may affect the internal operation of Sun. A correlation of astrophysical effects with results from current imaging techniques monitoring the operation of the Sun may provide important insights about the Sun and its relationship with the planets.

The identification of connections between the dynamics of the center-of-mass of the Solar System and its effects on the operation and/or structure of the Sun are neither new nor creative. Since the 1960's, several major studies elude to possible astrophysical connections[2]. During the 1990's quality software products were designed to report an approximate celestial location for the planets as a function of time. Often, the software products are presented in a book format with an option to purchase computer software disks for specific programs. **Astronomy With Your Personal Computer**[3], by Peter Duffett-Smith, is one such publication. This book presents computer programs in a modular format and designed to calculate planetary, orbital information (excluding Pluto) as a function of time. The programs were written in 'basic language' and with minor modifications one can design a different computational model, a computer program designed to report the approximate position for the Solar System's barycenter in time.

Extended, solar minima are times of minimal sunspot activity, low x-ray emission, and high solar radiancy. The Maunder minimum of ≈1642-1710 was an extended, solar minimum and it likely affected meteorological activity on Earth. Therefore, the time interval from 1550-2050 was selected for this study. Within the computational program an iterative interval of 15 days was used to compile the periodic sampling of orbital data. From this collection of data a refinement process for the final selection of data for analysis was defined. The selection criteria for the analysis included barycenter displacements at: 1) the beginning and mid-year, and 2) a maximum or a minimum. These data were transferred to the computer program[4] **Graphical Analysis** for further analysis and interpretation.

An Inferential Model : The Effect of A Dynamical, Planetary Center of Mass On Sunspot Activity

Abstract

Information and conjecture about the structure and operation of the Sun are prolific, but many truths about the Sun remain encoded within its radiancy. The Sun's orchestration of the planetary orbits is an example of a unique system existing in a state of equilibrium or near equilibrium. The Sun governs the motion and orbital configuration for each planet. And, the synchronous motion of the planets may temper and sustain some of the complex operations within the Sun. This equilibrium is fundamental to life and life forms existing on planet Earth.

Introduction

"And GOD said, Let there be lights in the firmament of the heaven to divide the day from the night, and let them be for signs, and for seasons, and for days, and years. And let them be for lights in the firmament of the heaven to give light upon the earth: and it was so. And GOD made two great lights; the greater light to rule the day, and the lesser light to rule the night: HE made the stars also."   Gen 1:14-16 (KJ)

The Solar System is an example of a vast, complex system in a state of un-ique balance. Each part of the system has its function, contributing to a precise structure and a harmonic operation of multiple, mechanical couplings. Describing the complicated interactions operative within the Solar System requires the design of elaborate models; models synthesized for creative questions, believable assumptions and approximations, and hypotheses which are consistent with describing the system.

As scientists study the Sun important questions about its composition, structure, and function lead to discoveries affecting the totality of science. Solar science has impacted the development of a standard model for matter, a model offering a strategy to unify the four operational forces (gravitation, electromagnetic, electroweak, and strong) in nature.

Validation and Consistency for a Computer Model

A scientific inquiry designed to monitor the motion of the Solar System's barycenter with respect to the center of the Sun is a logical entry point to begin this investigation. Each component alone the plane-of-the-ecliptic carries unique information. Each dimension (x, y, and z) may have different cause-and-effect relationships on the operation and stability of the Solar System. Identifying periodicities among the barycenter data and searching for connections with individual planets or combinations of planets is a reasonable objective.

This computer model was designed to provide output information in an iterative interval of 15 days. The results were tabulated in the following format:

- a radial location (x, y, and z) for the Solar System's barycenter
  > x and y components lie on the plane-of-the-ecliptic,
  > z component; perpendicular to the plane-of-the-ecliptic, and
  > radial, separation algorithm, $(x^2 + y^2 + z^2)^{1/2}$, and
- time in years (rounded to two decimals places).

The graphical analysis and interpretation for this investigation use a subset from this data. The sampling routine for this subset of data included the following data selections: the beginning of a year, mid-year, and any maximum or minimum, barycenter displacement.

While the exclusion of Pluto's contribution to this complex interaction likely has but a minimal effect, it remains that most models present only a glimpse of reality. Using reliable, numerical information at the 'input step' governs the success of a computer model. Credibility for a computational model is established by searching the 'output data' looking for identifiable data patterns and trends reported in earlier investigations. Fairbridge[5] and Shirley report common, astronomical periodicities within the Solar System with time intervals near 11, 60, 80, and 180 years. Also, the search looked for a correlation of the barycenter displacement and special events on Earth. Chief among these connections is a prolonged, solar minimum and its potential stimulation of meteorological events like those reported with the Wolf and Maunder minima. These guideposts are helpful in forming questions, hypotheses, and predictions for both past and forthcoming events.

Orbital Periodicity and Synodic Couplings for the Major Planets

The most significant orbital unit within the Solar System is the planet Jupiter. Jupiter's mass is less than that the mass predicted to form a binary system. But, the radial displacement for the center-of-mass for Jupiter and the Sun oscillates between 7.06E8m (slightly larger than one solar radius) at perihelion and 7.78E8m (about 1.1 solar radii) at aphelion. Orbital stability for a satellite depends on the barycenter falling within the structural radius of the primary body. Therefore, to maintain orbital stability within the Solar System, the center-of-mass should reside within the solar radius for a majority of time. A periodic transit outside the photosphere is possible, but within short time it must return to the solar interior and spend a majority of its time within the Sun. It is within the realm of possibility that orbital structure and dynamics and the operation of the Sun depend on the motion of the barycenter.

The Jupiter-Saturn syzygy (an alignment of the Sun and planets) produces the most significant gravitational effect among the planetary, orbital couplings. Following is a listing of sidereal, orbital and synodic periods (intervals of time between successive conjunctions or oppositions) for the major planets:

PLANET		GRAVITATIONAL COUPLING	SYNODIC PERIOD (Y)
Jupiter:	11.862	1) Saturn	19.86
		2) Uranus	13.81
		3) Neptune	12.78
Saturn:	29.458	1) Uranus	45.36
		2) Neptune	35.87
Uranus:	84.012	1) Neptune	171.38
Neptune	164.796		

Table 4.1: The orbital periods and synodic couplings for the major planets.

The Dynamics of the Solar System Barycenter

An examination of the radial displacement of the barycenter for the Solar

System with respect to the center of the Sun over a long interval of time should exhibit many of the common periodicities identified by astronomers. Each reported radial displacement has an **x** and **y** component on the plane-of-the-ecliptic and a **z** component acting perpendicular to this plane.

Special attention should be given to the 'trends and patterns' offered through the distribution of maxima and minima associated with the radial displacements. These sites may be associated with solar limits and/or mark the beginnings or ends for the significant periodicities. Following is the graphic representation for the radial displacement of the center-of-mass of the Solar System with respect to time.

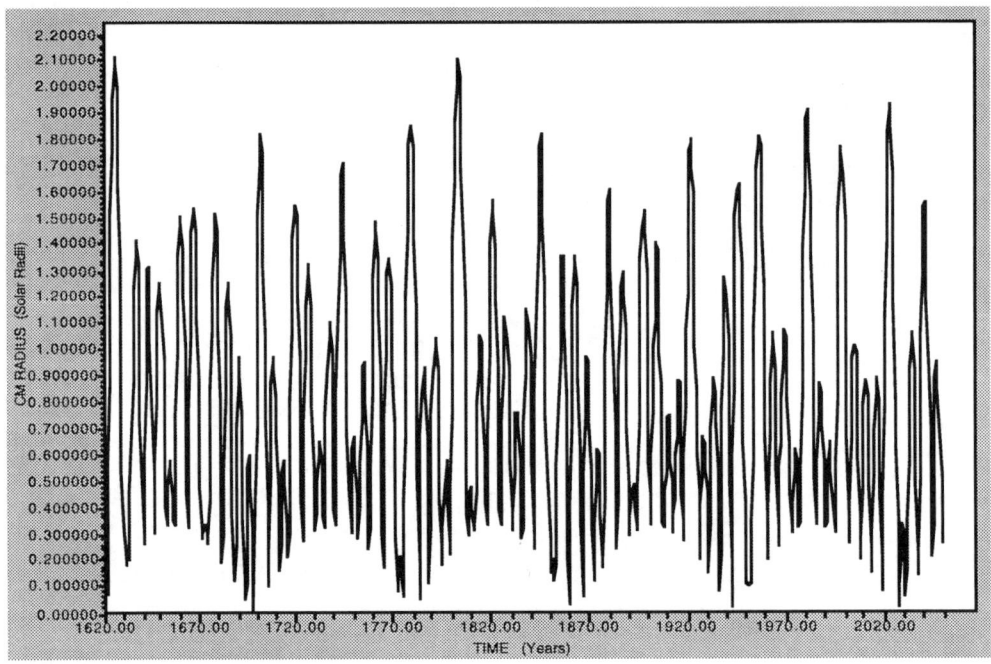

Graph 4.1: A display of the model's data showing a radial separation for the barycenter of the Solar System with respect to the center of the Sun, 1620-2050.

One pattern identified on this graph is the cyclic distribution found with the radial minima. A good geometric shape to associate with this pattern is a cycloid. During this 400y interval, an average periodicity for each cycloid is approximately 82.16 years. Following is a list of prominent, astronomical periods comparing favorably with this nearly 82y interval:

Orbital Event	Cycle	Interval
Jupiter period (11.86y)	$\cong$ seven, orbit cycles	7 X 11.86y = 83.02y
Jupiter-Saturn (19.86y)	$\cong$ four, synodic cycles	4 X 19.86y = 79.44y
Uranus period (84.01y)	one, orbital period	84.01y
Neptune period (164.8y)	$\cong$ two cycloid cycles	2 X 82.16y = 164.32y

Table 4.2: Astronomical connections to the cycloid pattern in **Graph 1**.

Another feature associated with the cycloid pattern is the ultra-minimal, displacement values of about 0.004 solar radii during the Maunder minimum (1645-1715) and the value near the turn of the 17th century, during 1698. Also noteworthy are the minimal value of about 0.02 solar radii appearing during the years 1616, 1942, and 2027. While these values may not represent limits, they may provide correlative information with emergent, structural models for the Sun.

A third, identifiable pattern is among the maxima displacements. The barycenter moved outward from the Sun to approximately 2.15 solar radii during 1625 and again during 1803. The interval of time between these maxima is approximately 178 years. (Remember, the synodic period for Uranus and Neptune is 171. 35 years.) Several recurrent patterns and periodicities display this approximately 178y period. Also, many of the amplitudes increase and diminish with time suggesting undulations indicative of a resonant, harmonic wave pattern; perhaps, a form of gravitational resonance.

To facilitate the identification and interpretation of some of the major events and periodicities during this four-century interval, examine **Graph 4.2** and its summary in **Table 4.3**. These major patterns include:

- maxima of 1625 and 1778 plus $\cong$ 20y (Jupiter-Saturn synopic period) brings a beginning for the Maunder minimum in 1645 and the Sabine event of 1798.

- periodicities: 59.24y, 82.16y, 177.7y, and 355.43y.

The interval between the maxima of 1625.56, 1803.29, and 1980.99 display the near 178y periods. Actually, this interval may form from three smaller cycles of 59.2 years. **Tables 4.8** and **4.9** suggest other events associated with this nearly 60y interval.

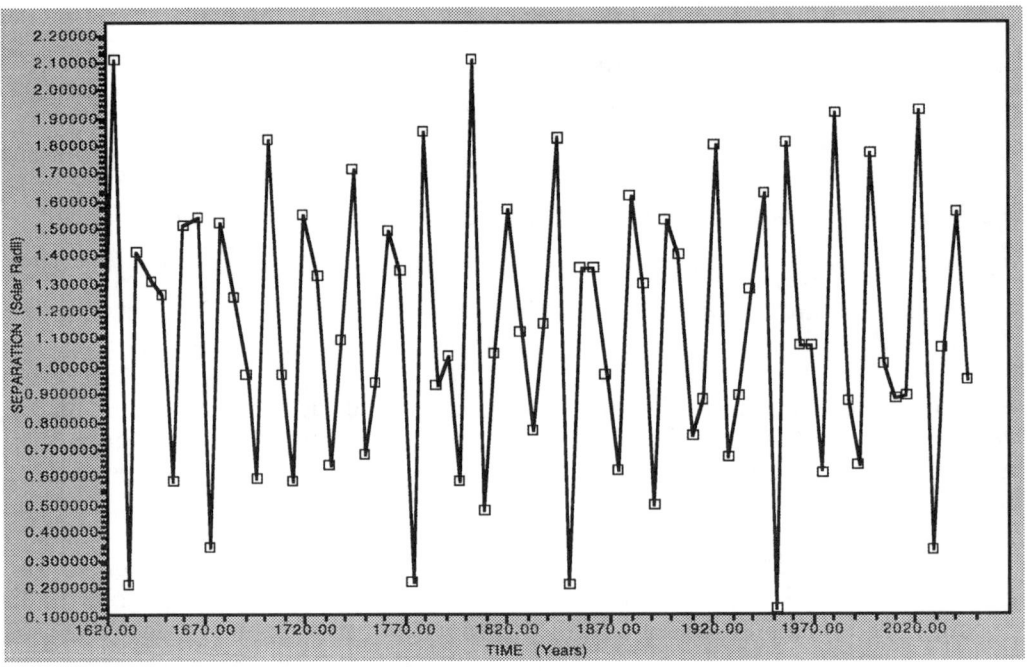

Graph 4.2: Maxima separations for the barycenter of the Solar System as a
function of time between 1620-2050.

TIME (Y)	SEPARATION (Solar Radii)	THREE CYCLE INTERVAL (Y)	TWO CYCLE INTERVAL (Y)
1625.56	1.3094	-	0.00
1684.83	1.2524	59.27	-
1744.11	1.7077	59.28	-
1803.29	2.1067	59.18	177.73
1862.63	1.3567	59.34	-
1921.85	1.8015	59.22	-
1980.99	1.9082	59.14	177.70

Table 4.3: A long period, 355.43 y composed of two cycles ($\approx 177.7^+$ y)
each and three shorter cycles (≈ 59.25 y) each, during the time
interval 1625-1981.

interval, 1620-2050.

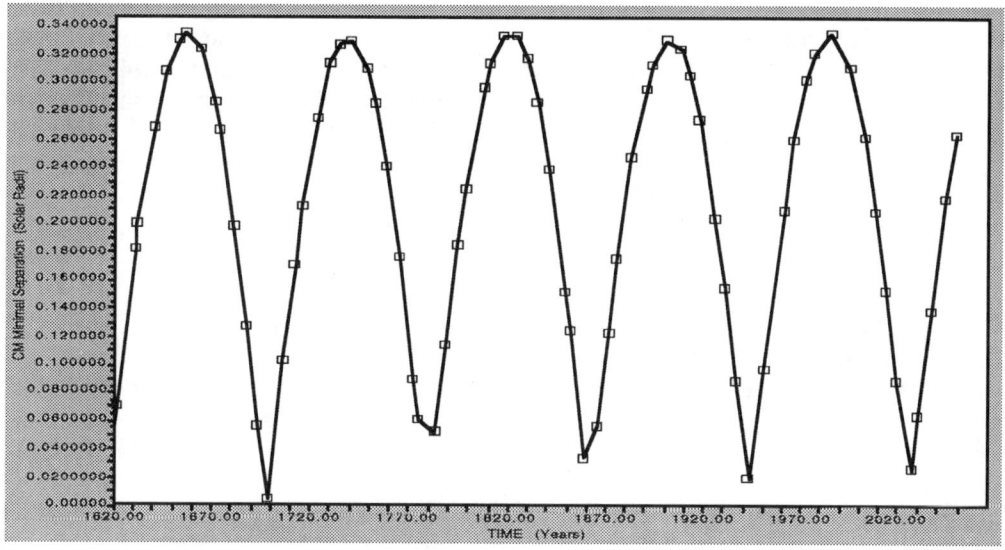

Graph 4.3: Minimal separations for the Solar System's barycenter, 1620-2050.

The periodicity for the minima is nearly 80y and is similar in pattern to the **y**-component noted in **Graph 4.6**. Values in **Table 4.4** link two of these cycles to form two larger interval periods of about 166 year (near the orbital period of Neptune at 164.8y), and an interval of near 410 years. The following data table displays this logic:

TIME (Y)	SEPARATION (Solar Radii)	INTERVAL (Y)	TWO CYCLE INTERVAL (Y)
1574.35	0.3308	-	-
1653.00	0.3310	78.65	0.00
1740.33	0.3303	87.33	-
1817.49	0.3336	83.94	165.98
1901.19	0.3310	83.70	-
1985.13	0.3361	83.94	167.64

Table 4.4: Five cycloid cycles, 410.78 years, maximum to maximum, beginning prior to the Maunder minimum and ending with the cycle before a **predicted** extended, solar minimum.

The approximate periods of 84 and 166 years compare favorably with the orbital periods of Uranus (84.012 y) and Neptune (164.796 y). Again, the influence of the major planets on the motion of the Solar System's barycenter is a very positive correlation. Because of synodic alignments and their periodic, gravitational impulse an important period in the approximate range of 410 years, or some multiple thereof, should be identifiable. **Table 4.5** summarizes this connection:

SYNODIC IMPULSE	SYNODIC PERIOD (Y)	INTEGER	INTERVAL (Y)
Jupiter-Saturn	19.86	21	417.06
Jupiter-Uranus	13.81	13	414.30
Jupiter-Neptune	12.78	32	408.96
Saturn-Uranus	45.36	9	408.24

Table 4.5: A resonant, synodic orbital configuration among the major planets having an average interval pattern of about 412.14 years.

Also, a connection with the maximum separation of the barycenter at 1980.99 and the time interval from 1999 to 2010 may signal a coming extended, solar minimum. Fairbridge and Shirley predicted a prolonged, solar minimum beginning as early as the 1990 cycle. That did not happen. But, they may have missed with their prediction by one sunspot cycle and that would align favorably with the prediction from this model.

> > A Predicted, Prolonged Solar Minimum, 2012± 11y

The last reported solar maximum was during 1990, 1990.1. If one adds the average sunspot interval of 11.1 years to the 1990.1, the next predicted maximum should occur during 2001. The National Oceanic and Atmospheric Administration (NOAA) in Boulder, Colorado is predicting a maximum number of sunspots (also, maximum x-ray emission) for this cycle at 160± 30 (phone conversation: October 1997). This is a large value when one examines the sunspot activity as a function of time for the past two or three centuries. An interpretation of the data in this model suggests a pattern beginning with the maximum separation of 1980.99 being similar in pattern to the maximum of 1625.56, the maximum just preceding the Maunder minimum of 1645. Adding 19.86 years, the synodic period for Jupiter and Saturn, to each of these dates marks a beginning for the Maunder minimum, 1645.42, and a possible beginning for an extended, solar minimum, 2000.85. The following table tabulates this correlation:

MINIMUM	YEAR	MAXIMUM SEPARATION (Solar Radii)	INTERVAL (Y)
Maunder:	1625.56	2.1120	0.00
	1631.48	0.2049	-
	1636.69	1.4099	-
	1643.18	1.3094	17.62
Predicted:	1980.99	1.9082	0.00
	1987.47	0.8731	-
	1992.48	0.6428	-
	1998.69	1.7672	17.70

Table 4.6: Adding 2.2 year to the two maxima of 1643.2 and 1998.7 is comparable to adding 19.86 years to the maxima of each 1625 and 1980. Both of these comparisons agree with the beginning of the Maunder minimum and a prediction for an extended, solar minimum beginning as early as 2001.

The Sabine event may have been associated with a weak, extended solar minimum. This event is purported to having an affect on Earth's weather pattern for about 25 years (1798-1821). This interval falls at the end of a 178y cycle, but does not contain the progression of maxima separations (three successive increasing maxima instead of three slightly diminishing maxima) as does the two cycles indicated in **Table 4.6**.

> Components of the Radius Vector for the Solar System's Barycenter, 1620-2050:

The orbital planes for each planet, except for Pluto at about 17 degrees, move close to the plane-of-the-ecliptic, a projection of the plane formed between the Sun and the Earth. Mercury is inclined at an angle of about seven degrees followed by Venus at 3.4 degrees and the other planets at a lesser inclination. Therefore, one could hypothesize the **x** and **y** components of the radial displacement carrying the significant trends, patterns, and periods to correlate with the motion of the Solar System's barycenter. The next series of graphs (**Graphs 4.4-4.6**) and data tables (**Tables 4.7-4.9**) focus on an analysis for each of these components.

> > The X-Component

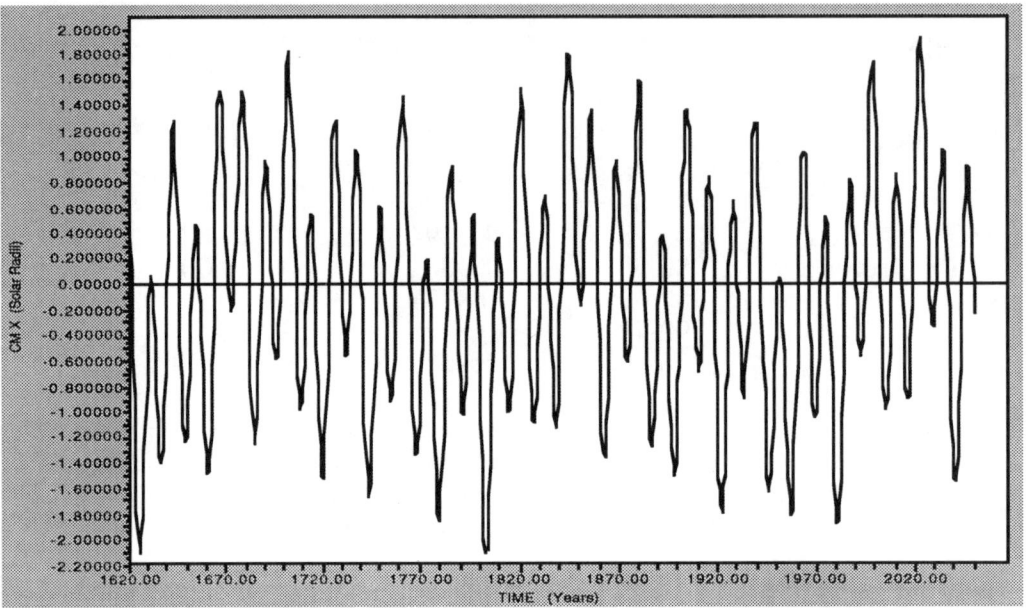

Graph 4.4: The x-component associated with the Solar System's barycenter radius, 1620-2050.

The x-component, right ascension, is defined to fall along the standard, astronomical axis defined by astronomers to identify objects in the Solar System. Right ascension (R.A.) is an angle formed on the plane-of-the-ecliptic with its increase moving in a counter-clockwise direction as viewed from above the plane. Zero degrees R.A. begins with a straight line drawn from the Sun and aligning with Earth's autumnal equinox. The y-component, like a Cartesian coordinate, is perpendicular to the x-component.

The primary resonant, contributions from the planets manifest their effect along the plane of the ecliptic. All of the periodicities identified with the analysis of the radius vector can be found among the x-displacements. Following are the major periods identified in the first three graphs (analysis of the barycenter radial displacement): 1) 11.85y, 2) 59.25y (5 X 11.85y), and 3) 177.75y (3 X 59.25y).

Following is a table of values summarizing the identification of these cycles on the x-component graph, **Graph 4.4.**

CYCLE (Y)	YEAR	SHORT INTERVAL (Y)	LONG INTERVAL (Y)
59.25	1702.38	0.00	0.00
	1761.52	59.14	-
	1820.94	59.42	-
	1880.12	59.18	177.74
	1939.38	59.26	-
	1998.69	59.31	-
	*2057.87	59.17	177.75
177.7+	1625.56	0.00	0.00
	1803.00	177.44	
	1980.99	177.70	355.43

Table 4.7: Major periods associated with the x-component.

> > The Y-Component

A graph of the **y**-component is the most sinuous of all the radial components. Its periodicity is approximately 166 years and follows in magnitude close to the orbital period of Neptune, 164.796 years. Half of the period approximates the 83y cycle of the cycloid in **Graph 4.1**. Another solar activity one can associate with the 83y period is the solar flare cycle. The positive **y**-component is on the plane-of-the-ecliptic and directionally near Earth's perihelion at about 103° right ascension.

There are some similarities between **Graph 4.5** and **Graph 4.3**, the graph displaying values for the barycenter, radial minima. The primary difference is the phase of the two graphs; they differ by approximately 180°. A primary similarity in the two graphs is the near 166y periods.

Following is **Graph 4.5**, the **y**-component for the Solar System's radius as a function of time:

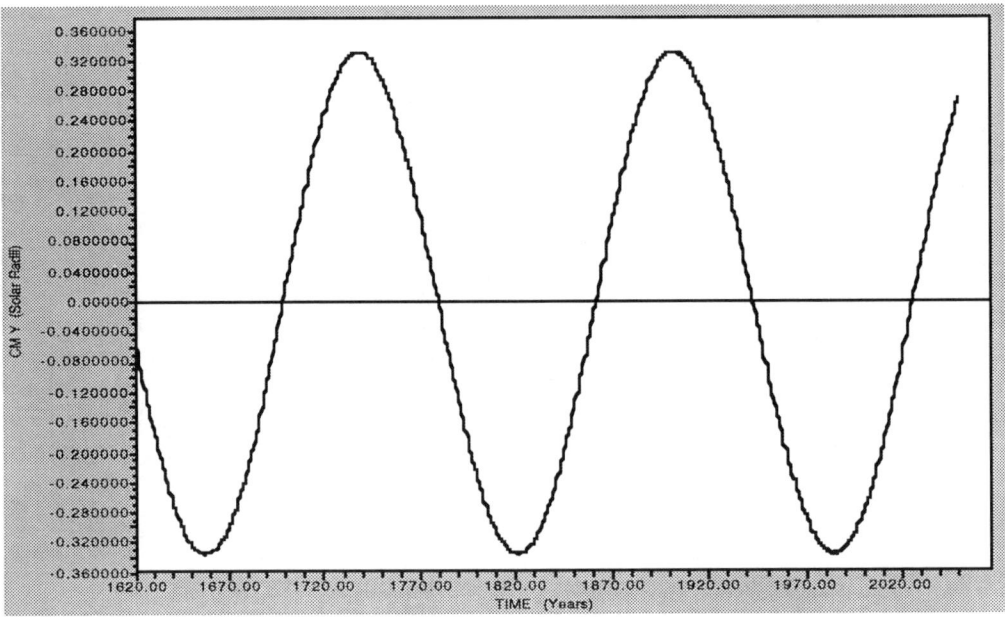

Graph 4.5: The **y**-component associated with the radius of the Solar System's barycenter, 1620-2050.

> > The Z-Component

The positive, **z**-component rises upward from the plane-of-the-ecliptic, the **x-y** plane. If one compares the graphs for the **z**-component and the **x**-component, two important trends can be identified: 1) all of the radial, periodicity patterns and 2) a near, complete phase shift for maxima and minima (similar to the **y**-component analysis). Or, a positive, **x**-amplitude is nearly time compatible with a negative, **z**-amplitude.

Also, the **z**-component falls along the same line-of-action used by astronomers and astrophysicists to define the vectorial orientation for orbital, angular momentum. Most of the planets have a significant proportion of their spin angular momentum aligned along this same **z**-axis. Following is a representation of the **z**-component for the Solar System's barycenter as a function of time from 1620-2050:

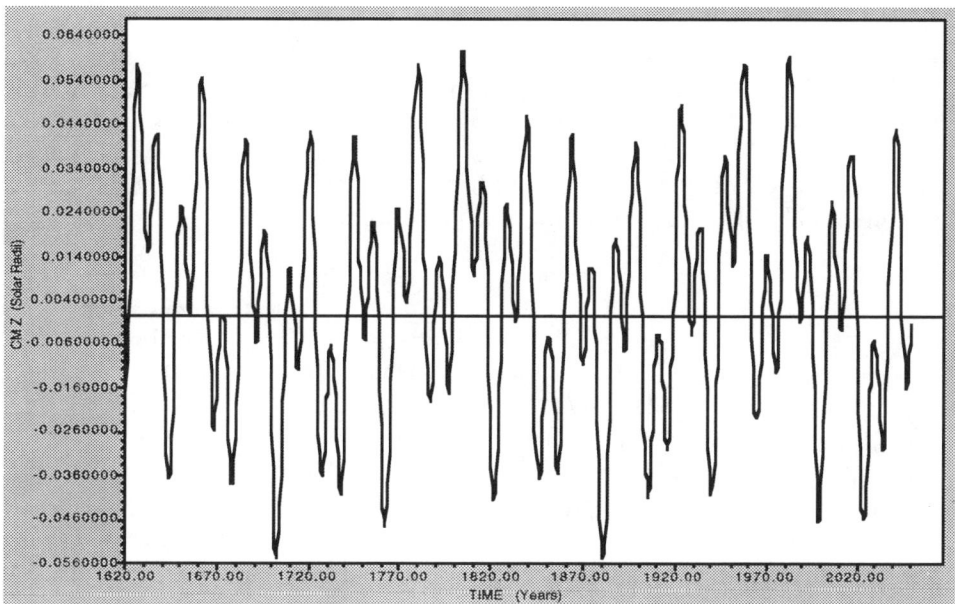

Graph 4.6: The z-component associated with the radius of the Solar System barycenter, 1620-2050.

CYCLE (Y)	YEAR	SHORT INTERVAL (Y)	LONG INTERVAL (Y)
59.25	1643.56 *	0.00	0.00 *
	1703.00	59.49	-
	1762.01	59.01	-
	1821.48 *	59.47	177.92 *
	1880.49	59.01	-
	1940.00	59.51	-
	1999.01 *	59.01	177.53 *
177.7+	1626.02	0.00	0.00
	1804.02	178.00	
	1981.48	177.46	355.46

Table 4.8: Major periods associated with the z-component.

(**Note:** The interval from 1643 to 1999 is about 355y and a common barycenter cycle.)

The periodicities associated with the radial components of the barycenter displacement provide evidence to predict a beginning for an extended, solar minimum. If the Maunder minimum began in 1645 and the Sabine event started in 1798, then the pattern would suggest another event beginning near the turn of this century. This 177.7y interval is very consistent in posting its signature within the analysis of data associated with the z-component of the barycenter displacement. **Tables 4.8 and 4.9** tabulate this trend.

Solar Event	Interval Date	Duration (yrs)	Interval from Ending (yrs)
Wolf minimum	1281-1347	66	-
Sporer minimum	1411-1524	113	177
Maunder minimum	1645-1712	67	188
Sabine event	1798-1823	25	111

Table 4.9: Events and intervals associated with a near 60y and 178y cycles.

Summary of Findings:

> A Hypothetical Cause for an Extended Solar Minimum

It is very presumptive to suggest any single event could cause such a significant effect as a two-decade, drought cycle. But, the record of this cycle may be associated with a timing mechanism of cumulative events within the Solar System. The periodicity of the Solar System's barycenter may be a primary contributor to the convergence of multiple events causing this ominous event. Following is a summary of the analysis for each of the major periods noted in this study:

Interval	Analysis	Effect
≅11.8 years	x-component graph z-component graph	≅23y magnetic cycle Jupiter orbital period
≅59.2 years	all radial graphs (except y-component)	five orbital periods for Jupiter
≅83.7 years	radius of CM	seven orbital periods for Jupiter cycloid period
≅162 years	y-component graph	orbital period for Neptune
≅177.7 years	all radial graphs (except y-component)	fifteen orbital periods for Jupiter
≅355 years	all radial graphs (except y-component)	30 orbital periods for Jupiter
≅412 years	resonant, harmonic for synodic configuration	35 orbital periods for Jupiter

Table 4.10: Common intervals and effects within the Solar System.

The model predicts the 'center of mass' penetrates to a minimal value of about 0.02 solar radii and a maximum extension of about 2.1 solar radii. In a stable orbital system, the barycenter should move within the radial boundary of the primary for predominate periods of time. And, the data from this model shows the center-of-mass spending a preponderance of its time within the photosphere. If the size of the Sun includes the 2.15 solar radii (about 1.5E9 meters), then the center-of-mass would spend all of its time within the Sun. But, that would require the definition of a new characteristic for the size of the Sun, a different solar radius.

The resonant, harmonic shape associated with these curves suggest the speed associated with the Solar System's barycenter is not of an orbital nature. Therefore, the tendencies associated with Keplerian descriptions should not be used to explain the motion associated with the speed of the barycenter with time. Also, calculations for this quantity indicate quantities exceeding the speed of light. The 'center of mass' is a virtual mass; therefore, it is not limited to the restrictions of mass transport in the realm dominated by the 'electromagnetic force.' It would seem the maximum speeds for the barycenter are near the edge of the photosphere and minimum values occur near the extremes of the displacement. The path of the barycenter may be in a spiraling loop like structure, with the loop not necessarily including the Sun's center, and moving between the amplitude limits near 0.02 and 2.15 solar radii.

If the barycenter moves from deep within the Sun to outside its photosphere, then this cycle may serve to agitate or stir matter within the Sun. The loop-like motion of the barycenter could produce a tornadic response to material within the Sun. It may stimulate some of the magnetic effects we associate with sunspot activity and the magnetic, reversal cycle. This same agitation may affect the rotational speed of the solar interior and may provide a mechanism to modify the neutrino flux coming from deep within the Sun. The periodicities associated with the dynamics of the barycenter could be a primary component influencing the internal operation of the Sun. Also, the minimal radial displacement of the 'center of mass' may define an operational interface; separating nuclear interactions from electromagnetic effects.

> An Imminent, Extended Solar Minimum

In 1987 Fairbridge and Shirley predicted an imminent, prolonged solar minimum to begin sometime during the time interval of 1990-2013. The result of this analysis and interpretation agrees with the most profound prediction from the Fairbridge-Shirley study, a forthcoming prolonged, solar minimum. This interpretation predicts the beginning of an extended, solar minimum during the interval 2001-2025.

Correlation with prior extended, solar minima suggest changes in normal meteorological cycles: a drought cycle in the southwestern United States and/or severe winters in northern Europe. If such an event were to happen at this time, it could have implications for world food production and a quality of living condition at various locations on Earth.

Near the year 1275, the Pueblo Indians residing at Mesa Verde, Colorado experienced a severe drought. Predictions suggest the drought may have lasted nearly two and one-half decades. Both domestic and agricultural water supplies became very limited. This drought state likely caused a southward migration of these cliff-dwellers and a search for a more, hospitable environment in which to survive. The Wolf minimum of 1281-1343 presents a positive, time correlation with the Pueblo Indian migration.

The 'Gabriel Cycle' is like a confluence of astronomical events. Planetary and lunar orbits combine to form an approximate 745y cycle. While consistent, positive correlative data to reinforce a periodic event on Earth with this cycle is lacking, adding 745 years to the 1276 date brings the date 2021 and additional evidence for the occurrence of an extended, solar minimum between 2001 and

2025.

Another algorithm can be applied to the data generated by this model. A 'trigger event' associated with the maximum separation of the 'center of mass' of the Solar System from the center of the Sun in 1625.56. Adding the synodic period for Jupiter-Saturn, 19.86 years, to this date brings the commonly reported beginning for the Maunder minimum, 1645. From 1625.56 to 1980.99, 355.43 years, two major cycles associated with the motion of the Solar System barycenter and a period of about 178 years transpired. This presents the maximum radial separation of 1980.99 as a recent 'trigger event'. Adding the interval time for the synodic period of Jupiter-Saturn, 19.86 years, to this value, brings a rationale for the 2001 and a possible date for another extended, solar minimum.

Lastly comes a common trend from the data. The y-component has a periodicity near 166 years, near the orbital period of Neptune. Adding one sunspot cycle to this interval yields the near 178 year cycle. Adding this 178y interval to 1645 yields 1823. The nearest, major minimal separation occurs in 1798.8, the purported beginning of the Sabine event. Adding 356y to the 1645 date brings the date 2001. The nearest, minimal separation occurs near 2002.4. Following is a data table summarizing this connection:

Date	radial separation (solar radii)	x-component (solar radii)	y-component (solar radii)	z-component (solar radii)
1646.22	0.3077	-0.0057	-0.3074	-0.0133
1798.81	0.2249	0.0043	-0.2248	-0.0039
2002.38	0.2617	0.0017	-0.2615	-0.0087

Table 4.11: Solar radial component separations for the 'center-of-mass.'

> An Argument For Gravitational Resonance Within the Solar System

The cycles associated with satellite, orbital motion within the Solar System is a compelling evidence for gravitational resonance. Among the primary evidence for this gravitational resonance is:

- the orbital effect of Jupiter, $T_o = 11.86y$, to astronomical processes within the Solar System,

- the effect of the gravitational impulse provided within the system at synodic periods, particularly the synodic period of Jupiter-Saturn at 19.86years,

  - a cluster of five, periodically changing amplitudes within a longer period of 59.25 years (3 X 19.86y = 59.58y),

• the manifest effect these motions exert on the Sun and the Earth.

Although Jupiter is too small to join with the Sun to form a binary system, our Solar System may be a classic example of a system operating under the influence of resonant, gravitational potential energy. A more detailed study of the geometry and energy options within the Solar System may provide evidence for the elusive 'gravitational wave', its characteristics, influences, and manifestations in nature.

The remainder of the book provides an argument for gravitational wave interference at work within the Solar System; perhaps, orchestrating the orbital motion of the planets, moons, asteroids, comets, and meteors within the influence of the Sun.

Chapter Four: References and Resources

1. Jose, P. (1965). The 179 Year Cycle of the Solar Inertial Motion. Astronomical Journal, 70, 193-199.

2. Brouwer, D., & Clemence, G., Methods of Celestial Mechanics. (New York: New York Academic Press, 1961).

3. Duffett-Smith, P., Astronomy With Your Personal Computer, (2nd Edition). (New York: Cambridge University Press, 1990).

4. Vernier Software. (1995), Graphical Analysis (2.0.3, 2nd Edition), [Computer Graphics Program]. Portland, Oregon: Vernier Software.

5. Fairbridge, R., & Shirley, J. (1987). Prolonged minima and the 179-yr cycle of the solar inertial motion. Solar Physics, 110, 191-220.

Others:

Phillips, K., Guide to the Sun, (Cambridge, Great Britain: Cambridge University Press, 1992).

Pasachoff, J. & Kutner, M., University Astronomy, (Philadelphia: W. B. Saunders Company, 1978).

Wagner, J., Introduction to the Solar System, (Philadelphia: Saunders College Publishing, 1991).

Chapter Five: Stepping Toward A Resonant, Gravitational Model For The Solar System

The Microcosm, The Terrestrial Realm, and The Macrocosm

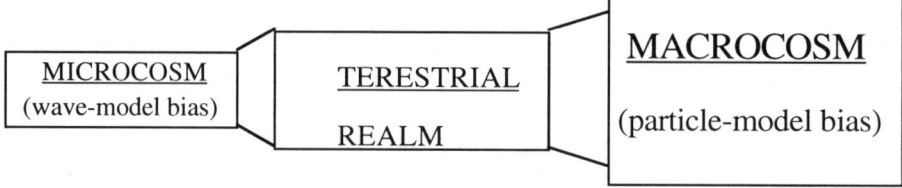

Figure 5.1: A model for the scaling span in science.

Figure 5.1 portrays one of the important tests for a scientific concept; its suc-cessful application across the scaling gamete of science, from microcosm to macrocosm. Successful modeling activities within one realm are often extrapolated and modified for trial within another realm. For example, the successes of seismology studies provide scientists with evidence of Earth's internal composition and structure. The inquiry techniques and analytical skills available to Earth scientists have been extended to solar scientists to study energy pulsations coming from the Sun. These inquiry strategies, a form of helioseismology, are designed to investigate the Sun's internal composition and structure.

The limitations of our sensorial systems greatly impede our ability to search for many, invisible truths about the universe. The foundation of the scientific inquiry process is built upon the gathering of credible evidence. With data, an inquiry process can step toward a successful interpretation for a natural event or process. Again, successes in interpretation and prediction with a modeling activity lead scientists to consider modifying the model and its findings to solve another scientific dilemma.

The atmosphere is a vital component for our existence; yet, its invisible state shroud many of the 'causes' provoking and/or moderating many of its 'effectual responses.' This cauldron of action and reaction present significant challenge to both the scientist trying to predict an atmospheric response for a given set of conditions and a public trying to respond to the prediction. The success of predicting meteorological events and/or hypothesizing about the 'cause and effect' of long-range climatic change stand as an impressive reminder of the order of complexity associated with a vast, dynamic system; a system with many interrelating variables.

Valuable information about the behavior and dynamics of fluids come from modeling activities using wind tunnels and water turbidity tubes. The identification of variables, the acquisition and analysis of data describing a response for these interrelating variables govern the formulation of an interpretation for the operation of a system. These descriptions for the behavior of a fluid under a set of conditions coupled with an log of historical experience provide scientists with evidence to design a mathematical model capable of predicting meteorological events and/or hypothesizing about imminent, climatic events.

Many of the complex interactions associated with atmospheric events require the use of probes to identify, define, and explore consistent responses from a primary or secondary event. Inference and hypothesis fuel these interactive, modeling activities; providing insight and evidence for scientists to form their interpretations for important and spectacular, natural phenomena. These scientific tools gather clues from remote and potentially hostile sites; providing observational evidence to analyze and describe states of matter and/or interactions beyond the range of our sensitivity.

Elemental hydrogen is not a natural state of matter, but science views elemental hydrogen as one of the simplest building blocks in nature. The scale for hydrogen's existence within the microcosm is beyond our innate, sensorial limits. Therefore, a model describing elemental hydrogen is dependent on credible evidence being gathered from well, designed investigations, inquiry strategies probing hydrogen's response within a controlled environment. These investigations provide feedback from observable, secondary events, perhaps an interaction producing a pulse of light or a reaction with another system component. A trail of consistent evidence and a successful prediction for a specific interaction with hydrogen tend to strengthen the model for elemental hydrogen; clarifying a view of hydrogen's structure and presenting a better understanding of its function in nature.

The distribution of energy surrounding the proton is among the structural interests for elemental hydrogen. Models incorporating a particle view, a wave view, and a wave-particle view have all contributed to a better understanding of hydrogen's response to different stimuli. The particle view, an electron orbiting the proton, is an analog of the orbital structures in the macrocosm. A form of Keplerian geometry can be applied to this view of hydrogen; presenting some interesting questions and shaping some engaging inferences about the similarities of these vastly different systems. If a unification pathway between the microcosm and macrocosm is possible, this view may become instrumental in initiating a connection.

The microcosm is a world of infinitesimal separations, small masses, and high, speed transfers. Probing the limits and operational conditions for energy states within the microcosm is imperative for scientific identification, definition, description, and interpretation. Any hope to present a unification of the microcosm with the macrocosm depends on the credibility of these definitions and reliable descriptions. For example, electromagnetic particles begin signaling of an impending transition, if not transformation, as they advance in translational speed beyond one-seventh light speed, $\cong c/7$. The excessive kinetic energy is transformed into forms of potential energy. Among these changes come an increase in mass and a dimensional contraction in the direction of propagation and/or an emission of radiant energy.

An electron's ground state, orbital speed may define another structural limit; a maximum for a stable, orbital speed, $v_1 = 2.1878E6$ m/s. This may be an ultimate, orbital speed for mass elements constrained within an electromagnetic field. The commonly occurring factor associated with this value, the fine structure (1/137.06) factor, manifest its signature in many of the mathematical models used to describe events and interactions within the world of atomicity.

Current operational models and interpretations for events within the microcosm lean toward a wave bias to account for the interactions at this diminutive scale. These models have been very successful in the provision of predictive hypotheses to account for inferred compositional and structural responses within the microcosm. The macrocosm presents an opposite bias; a bias toward a particle view. The interactive particles within the Solar System are large; differing in mass from the invisible 'particles' of the microcosm by approximately fifty 'orders-of-magnitude.' While the Sun and most of the planets are visible to the eye, the vast separations (about ten-to-twelve orders of magnitude from the Sun) pose another challenge for measurement and description.

Like the challenge for description in the microcosm, scale within the macrocosm may be the primary impediment challenging science to further unmask the geometry shrouding a more elusive role for waves and wave characteristics. Perhaps, a better balance of 'wave-particle' accountability within both of these realms will provide insight to form more comprehensive models; models designed to better characterize and predict nature's behavior.

Universal Constants in the Microcosm and Macrocosm

There are clear signs of similar, structural symmetries between interactions within the microcosm and within the macrocosm; particularly, among a particle view of the orbital systems within these realms. The two fundamental force relationships used by classical physics to describe interactions within these two Following is a graph and summary, data table for the minima during this areas have obvious differences; but, they also demonstrate compelling similarities in mathematical form:

Wave Bias with Electric Charge   **Particle Bias with Mass**

$$F_e = k \, q_1 \, q_2 / r^2$$
[electrostatic force]

$$F_g = G \, M_1 \, m_2 / r^2.$$
[gravitational force]

Figure 5.2: Classical expressions for the electrostatic and gravitational forces.

Both electric charge and gravitational mass use the concept of a field to account for the action of force through a separation. Both present a point, source geometry with a spherical, Gaussian field structure. Explanations for the interactions involving electric charge are formulated from a quantum view; a world governed by discrete, incremental quantities. Yet, helpful information comes from alternative system views. A particle view of electron orbital motion can facilitate a Keplerian argument for orbital motion within the microcosm being analogous to the model for planetary motion about the Sun.

The similarities of characteristics for orbital motion in both the microcosm and macrocosm, beg for a synthesis of a wave model to account for gravitational interactions. Again, a quantum model for gravitational interactions would necessarily encompass the classical view of gravitational. The symmetric connection between the classical relationships in **Figure 5.2** and the relationships describing gravitational interactions with a new model should include the following, analogous considerations:

- corresponding components within the Coulomb and Cavendish constants,

- a quantum view for mass corresponding to quantized electric charge, and

- a recognition of the point source geometry (Gaussian geometry) with quantized, radial separations, angular momenta, and total energy.

Both the Coulomb and Cavendish constants are classic examples for experimental constants. In 1785 Charles Coulomb's mathematical model describing the interactive nature for the electrostatic force included a value for his constant, $k=8.988E9$ [kg/C^2] [m^3/s^2]. Nearly a century earlier Newton calculated a value close to the modern, accepted value for the 'gravitation' constant, $G=6.672E-11$[m^3/s^2]/kg]. He proposed the Moon's orbit about Earth as an example of an object falling toward its central body. His logic (a mathematical model) encompassed the Keplerian constant and provided an explanation for the collection of constants forming the Keplerian constant, including Earth's gravitation constant, $g \cong 9.80$m/s^2.

In 1798 Cavendish used an experimental design similar to Coulomb's inquiry to better describe and evaluate the experimental result describing the gravitational interaction between two masses at various, finite separations. The investigative information provided Cavendish numerical evidence to confirm Newton's classical, mathematical model for gravitation force and calculate a plausible value for the Cavendish constant, $G = 6.672$ E-11 $[m^3/s^2]$ / kg.

> About the Coulomb Constant, k = 8.988E9 [kg/C²] [m³/s²]

An experimental constant usually is composed of other constants which describe the experimental environment and/or a unique collection and expression of universal constants. For example, the following four, different combinations of electromagnetic constants yield an appropriate value for Coulomb's, experimental constant:

    a. $k = [1/4\pi] [1/\varepsilon_o]$, where $\varepsilon_o = 8.85419$E-12 $[C^2/kg]$ $[s^2/m^3]$, the
        permittivity of 'free space' (vacuum) constant,

    b. $k = [m/e^2] [a_1 v_1^2]$, where the electron mass, $m = 9.1095$E-31 kg,
        $v_1 = 2.1878$E6 m/s, the Bohr orbital speed or the fine structure speed
        $v_1 = c / 137.03 \cong 2.1878$E6 m/s, $a_1 = 5.2918$E-11 m, the Bohr, orbital
        separation, and e $= 1.60219$E-19 C, elementary charge $(Ze=e^2)$,

    c. $k = [m/e^2] [c^2 r_e]$, where $m/e^2 = 3.5487$E7 kg/C^2, $c = 2.9979$E8 m/s,
        the speed of light, $r_e = 2.8179$E-15 m, the classical electron radius, and

    d. $k = [h/2\pi] [v_1/e^2]$, where $h = 6.6262$E-34 kg m^2/s, the Planck constant,
        and $h/2\pi = 1.0546$E-34 kg m^2/s, the Planck quantized, angular
        momentum constant.

Within numeric expressions, it is often difficult to see subtle, conceptual connections. But, if the idea of 'unification' is possible, then connections between the microcosm and microcosm must exist. One pathway should be the consequence of a connection between experimental constants associated with outcomes from inquiry models in each realm. Noting the unit associated with the Coulomb and Planck constants, permits the following connection to Keplerian geometry:

    • area per unit time (m^2/s) within Planck's constant, $h = 6.6262$E-34 kg m^2/s,
      and

    • separation cubed per period squared (m^3/s^2) associated with Coulomb's
      experimental constant.

Of the preceding four relationships, it is the final form, $k = [h/2\pi] [v_1/e^2]$, which may provide an important connection with the Bohr model and the synthesis of quantum, gravitational model. Also, this collection of constants, $k e^2 =$ 2.3072E-28 kg m^3/ s^2 = $[h/2\pi]$ v_1 and the value associated with the $r^3/T^2 =$ 6.4137 m^3/s^2 for the electron in various energy states in the hydrogen atom, form relationships to calculate a value for the mass of an electron:

$$m_e = k e^2 /[r^3/T^2]/ 4\pi^2 = k e^2 /[4\pi^2 r^3/T^2] = k e^2 /[r v^2] = 9.11E\text{-}31 \text{ kg}.$$

> > About The Gravitation Constant, $G = 6.672E\text{-}11 [m^3/s^2 / kg]$

In classical physics 'G' is often called a universal or fundamental constant. A universal constant infers a discrete, fundamental value with no component parts. One view of a universal constant holds it to be a scientific analog to a mathematical, prime number. A **new model** for gravitational interactions could propose the universal, gravitation constant to be an experimental constant. An analogous development for gravitation, similar to the associations made with the Coulomb and Planck constants and Keplerian geometry, might prove worthy of consideration.

Three mathematical associations orchestrate the connection between the gravitation constant and Keplerian geometry:

• the similarity of each constant in the force expression, electrostatic and gravitational forces,

• the inclusion of the unit element, **m^3/s^2** (an orbital quantity associated with Kepler's third, orbital principle), and

• **v$_1$**, a maximum orbital speed for particles within the electromagnetic realm, **v$_1$ = 2.18E6 m/s**. This comparison assumes the following representation:

$k = 8.988E9 [kg/C^2] [m^3/s^2]$ $G = 6.672E\text{-}11 [m^3/s^2]/kg$

$k = [h/2\pi] [v_1/e^2]$ $G = [h/2\pi] [v_1/m_1^2].$

A new quantity, an orbital mass element **m$_1$**, surfaces in this comparison. An interpretation for this quantity of mass is open to speculation and conjecture. How are the limits for an orbital, mass element dependent on the characteristics of the 'central body?' Could tidal forces (the gravitational force producing surfacial distortions on interacting, large masses) have any dependency on this value? Understanding the significance of a fundamental, orbital mass element could carry information capable of illuminating a pathway to begin forming a quantum view for gravitation. The following calculations are examples for this logic:

• as an orbital element,

$m_1^2 = [h/2\pi] [v_1]/G = 2.3072E-28 [m^3/s^2]$ kg $/ 6.672E-11[m^3/s^2]/kg$

$m_1^2 = 3.4582E-18$ kg^2 with the mass element, $m_1 = 1.8596E-9$ kg and

$m_1 c^2 = 1.6713E8J/1.60219E-19J/ev = 1.0431E27ev = 1.0431E15$ Gev.

• as an exchange particle,

$m_0^2 = [h/2\pi] c/G = 3.3616E-26 [m^3/s^2]$ kg $/ 6.672E-11[m^3/s^2]/kg$

$m_0^2 = 4.7386E-16kg^2$ with the mass element, $m_0 = 2.1768E-8kg$ and

$m_0c^2 = 1.9564E9J / 1.60219E-19J/ev = 1.2211E28ev = 1.2211E16$ Gev.

Just as the unification of the nuclear weak force and electromagnetic force defined a specific family of interrelated, exchange particles: W^+, W^-, Z^0, and the photon, the unification of the strong nuclear force and the gravitational force suggest the probability of another family of interrelated, exchange particles. This energy state may provide conditions for the existence of the elusive, magnetic monopole and its minimal, energy threshold, an energy environment on the order of 10^{17} GeV[1].

A Review of Bohr's, Particle Model of a Hydrogen Atom

> Assumptions Associated with a Classical View

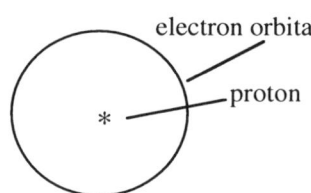

• Electromagnetic Force, $F_e = k Ze / r^2$

• Centripetal Force, $F_c = mv^2 / r$

• Orbital Energy:
   1) kinetic energy, $E_k = | k eZ / 2r |$,
   2) potential energy, $E_p = k eZ / r$, and
   3) total energy, $E_t = k eZ / 2r$
k - Coulomb constant, $k= 8.988E9 [kg/C^2] [m^3/s^2]$,   Z - proton charge,

   $Z = +1.602E-19$ C,   e - electron charge, $e = -1.602E-19$ C, with

   m - electron mass, $m = 9.10E-31$ kg,   v - electron, orbital speed,

   r - electron, orbital separation, and p - electron, linear momentum.

> Classical Conceptual Relationships Describing the Electron, Orbital Motion

   F_e (electromagnetic force) $= F_c$ (centripetal force)

$$F_e \quad = \quad F_c$$

   1) $k Ze / r^2 \quad = \quad mv^2 / r$,

   2) $E_k = |k Ze / 2r| = mv^2 / 2 = p^2 / 2m$, an expression for orbital, kinetic energy,

3) $p = e \, [km/r]^{1/2}$, an expression for electron, linear momentum, and

4) $E_p = -k \, e^2 / r$, with respect to the proton, this quantity represents the work necessary to bring an unbound electron of infinite separation to an orbital separation of radius 'r.' Notice the negative sign associated with this quantity. This signifies the binding nature of the potential energy for the 'electron-proton' system and it identifies the influence (opposite charges) causing the attractive, electrostatic force. Therefore, the total, orbital energy, E_t (for the electron with respect to the proton) and the electron, orbital angular momentum is given by the following relationships:

5) $E_t = E_k + E_p = |k \, Ze / 2r| + (-k \, e^2 / r) = - k \, Ze / 2r$, and

6) $L = r \times p$ electron, orbital angular momentum.

> **Assumptions Associated with a Wave-Particle View**

1) specific, stable electron orbitals (with quantized characteristics),

2) Keplerian geometry for an electron, its orbital path (a parsimonic view for the circumference of a circle) described by constructive interference, $C = 2\pi \, r$, or $2\pi r = n \, \lambda$, where **n** is an integer associated with a principle, quantum number and 'λ' is the electron's wavelength,

3) the electron angular momentum and total orbital energy are also quantized values, and

4) the energy descriptor for the electron's particle nature is given by the Einstein relationship, $E = m \, c^2$ and the electron's wave nature is given by Planck's relationship, $E = h \, \upsilon$.

> **Wave-Particle Relationships Describing Electron, Orbital Motion**

1) wave-particle duality, $E_p = E_w$ or $m c^2 = h \upsilon$, with $\upsilon = c / \lambda$, with $p = mc = h / \lambda$,

2) quantized orbital radii, $r = n \, \lambda / [2 \, \pi]$,   where n = 1, 2 or 3...,

3) quantized orbital angular momentum, $L = r \times p = [n \, \lambda / (2 \, \pi)] \, [h / \lambda]$ and $L = n \, h / (2 \, \pi)$, where the quantity '$h / (2 \, \pi)$', is a recurrent constant whose value is defined as 'h_{bar}', $h_{bar} = h / (2 \, \pi)$,

4) quantized orbital energy, $E_n = - E_k = - p^2 / (2m) = - [h^2 / \lambda^2] / (2m)$, equating the relationships for classical energy and wave-particle energy yields a result which can be associated with two important values:

a) orbital kinetic energy, $E_k = [h^2 / \lambda^2] / (2m) = |k \, Ze / 2r|$ with

$$\lambda = 2 \pi r / n, \text{ and } r_n = n^2 h_{bar}^2 / [m k e^2]$$

b) **if**, values for the collection of constants described in 'part a' are
given as follows: n = 1,   m = 9.10E-31 kg,
h = 6.626E-34 kg m^2 / s,   h$_{bar}$ = 1.0546E-34 kg m^2 / s,
k = 8.988E9 [kg / C^2] [m^3 / s^2], and e = -1.602E-19 C,

then, the value for the Bohr orbital, with n = 1, is given by
calculation to be, r$_1$ = 5.29E-11 m, and

c) the collection of constants other than the integer values for **n** define a
constant, the Bohr radius, $r_1 = h_{bar}^2 / [m k e^2] = 0.529$ A (Angstrom
units) with higher energy states defined by orbital radii, $r_n = r_1 n^2$.
The natural, upper limit for the 'integer n' is the number eight,
consistent with the eight periods associated with the periodic table.

5) total orbital energy, $E_n = - k e^2 / [2 r_n] = - k e^2 / [2 r_1] [1 / n^2]$. When the
principle quantum number, **n** = 1, the Rydberg constant is formed. The
value for the Rydberg constant is, $E_1 = - 2.188E18$ J = -13.59 ev. And,
the energy profile for an electron in orbital motion about a proton is
given by:

$$E_n = E_1 [1 / n^2] = - 2.188E18 \text{ J} / n^2.$$

The Bohr model of 1913 provided many insights and connections for
laboratory investigations with elemental hydrogen. In 1923 Louis deBroglie
proposed a modification for the Bohr model. deBroglie proposed a union for
the wave and particle models, a 'wave-particle' view for the microcosm. In
1926, Erwin Schrodinger's wave model replaced the Bohr model as a reliable
model to describe the structural geometry associated with the world of atoms,
the microcosm. Schrodinger's model was and is credited with an ability to
predict a broad range of experimental results for hydrogen and the other
elements.

Bohr's model for elemental hydrogen fell short of the mark in the arena of
scientific inquiry. But, the Bohr model provided science with an excellent,
opening step into the world of the microcosm; a beginning logic to question and
survey nature at a new, diminutive scale. Physical scientists had a new footing
and a new window to begin probing the mysteries of atomic structure; searching
for new 'patterns and trends' among earlier, experimental work and hypothesizing about new, inquiry models. New questions and hypotheses laid a
foundation for new, experimental results, establishing a platform for the formation of a comprehensive, quantum theory for electromagnetic interactions.

> An Inference from the Bohr Model

The Bohr model was built on the logic of Coulomb's electromagnetic force. Atomic hydrogen is a dynamic electric dipole. The calculational components of this model began to yield numeric values describing the interaction for the electron-proton coupling. Scientists had a new example to demonstrate the characteristics and responses from a very different orbital system. Among these calculated quantities were values for orbital characteristics: radial separation, speed, angular momentum, wavelength, and total energy.

Bohr used the Sun and planets as an analog for his early, geometric view of a hydrogen atom. Therefore, using the Bohr model as an analog for a gravitational interaction might prove a useful platform to synthesize questions and hypotheses for heavenly motions. The following assumptions from Bohr's model will be referenced in the development of a resonant, gravitational model in Chapter six:

a. wavelength is inversely proportional to orbital momentum ($\lambda = h / mv$),

b. integer, standing wavelengths ($n\lambda_1 = n * 3.3249E\text{-}10$ m) define the orbital circumference,

c. integer values for orbital, angular momentum, where $L_n = nh/2\pi$ and

$$L_1 = h/2\pi = 1.0544E\text{-}34 \text{ kg m}^2 \text{ /s,}$$

d. quantized space, $r_n = r_1 n^2$ with $r_1 = 5.2918E\text{-}11$ m and other associated values: $m_e = 9.11E\text{-}31$ kg $L_1/m_e = r_1 v_1 = 2A/T = 1.1577E\text{-}4 \text{ m}^2\text{/s,}$

$T_1 = 1.5198E\text{-}16$ sec, $T_n = T_1 / n^3$, $v_n = v_1 / n$, where

$$v_1 = 2.1878E6 \text{ m/s} = c / 137.036, \text{ and}$$

e. other orbital parameters to consider:

$k_{ss} = r^3 / T^2 = 6.4156 \text{ m}^3/ \text{s}^2$ $k_{ss} = v^2 r /(4\pi^2) = 2A/T (v_1 /4\pi^2),$

$E_n = [(\text{-}m_0 v_1^2)/ 2] [1/n^2]$ $L_n / E_{tn} = n^3 [T_1 / \pi].$

If the Bohr orbital is a fundamental, resonant state for elemental hydrogen, then the frequency of the electron in its minimal energy state may be in harmony with a fundamental frequency associated with elementary charge. Certainly this association is conjecture; but, if electric charge is frequency dependent, this would add another rationale to the successes of the Schrodinger wave function in modeling interactions within the microcosm. Lastly, if the atom is a resonant, electric dipole, then electron, orbital frequency, $f = 6.5833E15$ Hertz, may be associated with the frequency of a pulsating elementary charge.

The Bohr orbital frequency, **f = 6.5833E15 Hertz**, may not be the fundamental frequency associated with an elementary charge. This frequency may be a harmonic multiple of the fundamental frequency. Note the following:

$$E = h \upsilon = 6.626E\text{-}34 \ kgm^2/s \ [6.5833E15 \ Hz] = \text{-}4.36E\text{-}18 \ Joules$$

This value is approximately twice the energy associated with the Rydberg energy (-2.18E-18 Joules), the electron energy in the Bohr orbital. Perhaps, a more appropriate, fundamental frequency to associate with the elementary charge is the frequency, **f = 3.2917E15 Hertz**.

A Model for a Resonant, Gravitational System

Cosmologists use the scientific, conceptual framework as a guidepost to search the cosmos for signatures describing its composition, structure, and function. The principle of equilibrium, whether a static or dynamic equilibrium, influence the search for cosmological evidence. Equilibrium implies a balance or symmetry for logic. The principle of equilibrium is a footing to frame a question or form a hypothesis, perhaps a consideration for an observance and/or a proposal for a hidden structure.

The twentieth century has witnessed Albert Einstein's synthesis of the theory of relativity. This view of nature incorporated time into the traditional components of space, the dimensions of 'x, y, and z.' The 'Relativity Theory' specifies that all energy transitions and/or transformations occur on the 'space-time continuum'; that, the known universe is on an expanding 'space-time continuum.' Observational evidence and critical thought about the nature of a 'space-time continuum' lead cosmologists to form creative models to describe this invisible entity. Questions, hypotheses, and predictions about its structure and function are influenced, if not governed, by the principle of equilibrium.

The energy density for this structural paradigm seems to demand constancy. The formation of new matter must be offset by the transformation of older matter into energy. A localized event may signal a concentration of energy, an 'energy high,' but close by there should be an 'energy low.' Symmetry and/or equilibrium infer the existence of one depends on the other. Like the dynamics an interacting 'high' and 'low' pressure cell in a weather system, it is probable that chaotic vortices of energy result during the collision of a cosmic 'high' and 'low.'

The distribution and dynamics of mass elements within the Solar System may map the curvature and vorticular characteristics of space-time associated with the Sun. The window of observance for the Solar System is so short compared with the span of creation's time that it is permissible to assume this system to be nearly closed system. The gravitational low, the negative distortion of space-time close to the Sun, has been measured and found to be in accord with the result from the calculational framework within the theory of relativity. Certainly other energy lows manifest their presence within the Solar System as satellites orbit about several of the planets. These visible, gravitational lows beg for a question, a hypothesis, and an inquiry. Asking the question is the easy part: Are any gravitational 'highs' present within the structure of the Solar System? Identifying, probing, characterizing, and ... a 'gravitational high'; now, that is or will be a significant, scientific challenge.

If gravitational 'highs and lows' exist, then a tighter argument for analogous structure can be made for correlation of the Keplerian view of Solar System and the particle view for elemental hydrogen. Earlier, a conjecture about the nature of electric charge having a frequency, dependent component was inferred. If positive and negative, elementary charge are in different, frequency phase states, then a modified, geometric view suggesting how these charges distort their 'space-time continuum' would be a nature consequence. A positive charge will continue to manifest a 'hill-structure' distortion of the continuum, a state of high, electric potential energy; yet, a new provision for a pulsating structure should be added. Like the effects of an electric dipole, a pulsating 'source and sink' structure, a set of pulsating gravitational 'dipoles', should affect the geometry and interpretation of gravitational interactions within the Solar System.

Remember, the asymmetric statement of Kepler's first, orbital principle: 'All planetary orbits are elliptical with the Sun at one focus.' Among the assumptions for a resonant, gravitational model, could be an association of the empty set in an elliptical orbit and an invisible point of high, gravitational potential energy. If this assumption incorporates a provision for a periodic pulsation, then logic for balance and dynamics along the 'space-time continuum' become an interesting possibility. Any mass element moving slowly into the spacial proximity of a point of high, gravitational potential energy would be deflected from this location.

A wave interference model for gravitation must present a new or modified geometry for gravitational interactions. As the inquiry proceeds, the development of a hypothesis modifying the asymmetric, Keplerian geometry restores a nearly perfect symmetry for an elliptical orbit. The development of a model for gravitational interactions (Chapter Six) will use a traditional framework for orbital motion in the Solar System given by Kepler and Newton in the seventeenth century; but, a proposed model adds a logic for gravitational wave interference from two, point sources, pulsating 'out-of-phase.'

Summary and Speculation

1) If the Solar System is a resonant, gravitational system, then the classical, Keplerian geometry used to describe planetary motion requires a modification. Following is a list of assumptions for a resonant, gravitational system:

- the energy state for a stable orbit would likely be less than the classical interpretation,

- analogous, structural characteristics with a particle view for satellite motion within the Solar System and the electron orbitals of the Bohr model,

- a more, comprehensive view for the partition of energy within a gravitational system like the Solar System.

2) Many of the constants associated with the interpretation of events within the microcosm are really experimental constants. Chief among these constants is Planck's constant. Expressions for this constant and its alternate form, $h_{bar} = h/2\pi$, follows:

- $h = 6.626E\text{-}34 \ kgm^2/s = h = m_e \ v_1 \ \lambda_1$

  $h = [9.1095E\text{-}31 \ kg] \ [2.1878E6 \ m/s] \ [3.3249E\text{-}10 \ m]$

  $= 6.626E\text{-}34 \ kgm^2/s$, and

- $h_{bar} = h/2\pi = m_e \ [2A/T_0] = 9.1095E\text{-}31 \ kg \ [5.7876E\text{-}5 \ m^2/s]$

  $= 1.056E\text{-}34 \ kgm^2/s$.

3) Trying to establish an analogous geometry for the microcosm and the macrocosm is subtle business. A linking geometry, perhaps an orbital geometry, could provide a window to view gravitational interactions with a wave model. Clues must be present, but their manifestations are masked. For example, molecular hydrogen has two, common covalent states. These states have different energies by virtue of different electron, spin angular momenta. When the spin state for the two electrons is parallel in alignment (rather than antiparallel), the molecule is in a higher, energy state. The following diagram[2] represents the electron probability for these two different energy states:

parallel, electron spins:　　　　　**antiparallel, electron spins:**
(high, electromagnetic energy state)　　(low, electromagnetic energy state)

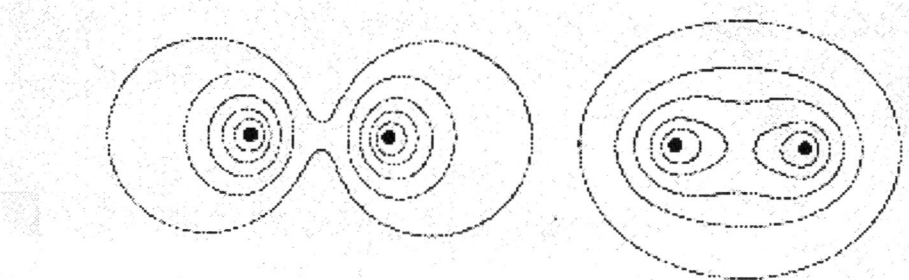

Figure 5.3:   Contours of electron probability for two adjacent hydrogen atoms whose electron spin states are antiparallel and parallel. From *Concepts of Modern Physics* by A. Biser , 1967 with permission from The McGraw-Hill Companies.

The variation of potential energy as a function of atomic separation for a covalent, hydrogen molecule with an antiparallel, electron spins state decreases to a minimum (maximum binding energy) when the proton separation nears 0.74 Angstrom units[3]. This field representation of antiparallel, electron spin state is similar to the potential energy model associated with lattice energy[4] in a sodium chloride complex and other ionic structures. Therefore, the geometry associated with the electric field in the antiparallel, electron spin state of a hydrogen molecule is analogous in structure to a classical, Coulomb view.

This molecular system is among the most simple in the microcosm. Any orbital similarities between these electromagnetic couplings and the orbital configurations of the macrocosm shape the assumptions necessary to form a wave, gravitational model. Certainly the Solar System is not a binary system with two stars and there are not two planets moving about a binary in an elliptical orbit. But, an adjustment of the components associated with a planetary orbital might prove beneficial. An inference for a planetary orbital might include: 1) the Sun at one focus, 2) an antiparallel, spin-angular momentum structure at the other focus (a point of high gravitational potential energy), and 3) a planet moving serenely in its elliptical orbit.

4) Within large stars, all four known force fields may coexist simultaneously. The structure and composition of a star is dependent on the confinement energies at each transitional state: gravitational to nuclear (both weak and strong interactions) and nuclear to electromagnetic. As energy moves from one confinement zone to another, the energy profile within that zone provide signatures of transformation. These changes may include: 1) a fissure into new particles, 2) a new set of boundary conditions for propagation (length, mass, and time), and 3) a confinement pattern with differing degrees of freedom to express its new energy state.

Electromagnetic interactions tend to dominate our realm of experience. In this region of the universe electromagnetic energy has the least confinement and the greatest degree of freedom. The primary exchange particle in this realm is the photon, a quantum of electromagnetic energy. So, a photon should propagate at greatest speed, experience the longest time periods (slow clocks), and be in a state of least energy. As photons pass from air to water, their speed of propagation is slowed and their wavelength is shortened. The photon begins to experience a mode of confinement. And, so it would be at each of the interfaces defining an operational force: electromagnetic, weak nuclear, strong nuclear, and gravitational.

High, energy colliders offer evidence for a different world, a unique environment formed by mutual couplings, transformations, and wave-particle characteristics. These artificial environments provide insight to infer pathways for energy transformation; clues of structure and composition on both sides of an interface dominated with different forces. If gravitational confinement is characterized by high energy [large elementary particles, small wavelengths, long time intervals (slow clocks), etc.], then the electromagnetic realm must be a setting for least, energy states.

As the exquisite detail of orbitals in the atomic setting is understood, it may provide clues to describe analogous interactions and structures in the vast expanses of the macrocosm, events orchestrated by the gravitational force. New modeling activities for gravitational interactions will unveil the true nature of gravitation, perhaps revealing implicit connections at each interface between nature's operative forces. These connections should clarify a unification among natures forces.

5) 'K-capture' is an interaction between an electron and an atomic nucleus. The electron is thought to spiral from its orbital configuration to greater speeds, experiencing an alteration of characteristics to become compatible with a proton cross-section, and an eventual collision/interaction with the proton. The concluding signal for this transformation is the formation of a neutron as modeled by beta-decay.

The manifest effects associated with an electron being accelerated in a cyclotron to higher and higher energies, would be analogous to this event.

Change in the particle's electromagnetic characteristics herald a forthcoming transition into a nuclear realm and a new set of boundary conditions. This beta-decay mechanism results in the formation of a neutron and the emission of gamma radiation and a neutrino. The beta-decay model for the capture of an inner electron by the rubidium-81 nucleus produces krypton-81, an anti-electron, and a neutrino.

Large stars, about two to three solar masses, convert all of their protons to neutrons in short time, $\cong$3-5 seconds, by a 'K-capture' mechanism. The energy and neutrino release are but two characteristics of this category of supernova. 1987-A presented science with a spectacular event for observation and intense study. Among the earliest evidence from this event was a neutrino burst arriving slightly before a trailing light burst. Many discoveries have offered science insights into the process of gravitational collapse. Chief among the discoveries associated with this event was the identification of a new state of matter, a rapidly rotating, neutron star.

References and Resources for Chapter 5:

1. Lederman, L. & Schramm, D., <u>From Quarks to the Cosmos</u>, (New York: W. H. Freeman and Company, 1995), p 172.

2. Beiser, A., <u>Concepts of Modern Physics</u>, (New York: McGraw-Hill Companies, 1967), p 223.

3. ibid, p 223.

4. ibid, p 219.

Chapter Six:
A Gravitational Wave, Interference
Model for the Solar System

Geometry is a window through which creative physics must pass. Once the geometry associated with a spacial problem is envisioned, the interactive process of questioning and hypothesizing about an associated problem can begin. Scientists must make connections between 'real world' components and intellectual assumptions, conjectures, and/or hypotheses to be used in an interrogation process. Johannes Kepler struggled for years before seeing a comprehensive geometry to connect with planetary orbits. Without his geometric insight, his formulation of the three principles describing planetary orbits may not have been synthesized. Isaac Newton used the results of Kepler's work to fashion his view of gravitation both terrestrially and within the heavenly realms. Newton offered a cause for Kepler's effect, a comprehensive description for satellites in orbit about a primary mass. Newton's comprehensive, cause-and-effect interpretation laid the foundation to explain parabolic trajectories near the Earth's surface, elliptical trajectories for the heavenly orbs, and became the pillar for the formation of his theory of mechanics. This is a classic example of two scientists working in tandem to provide a description and an explanation for a physical event.

Kepler was able to connect the geometry of an ellipse with the shape of planetary orbits. Mars, being close to Earth and having a significant orbital eccentricity, e = 0.092, was an excellent choice for observation and analysis. Mars' eccentricity was large enough to clearly delineate its elliptical shape from a circle. The agreement between the mathematics of an ellipse and Tycho Brahe's observational data for Mars encouraged Kepler to make his next, significant connection; the formation of a hypothesis suggesting all planets move in elliptical orbits. Armed with these comprehensive connections, Kepler began forming a geometric model for satellite motion within the 17th century cosmos. His operational assumptions and hypotheses provided a foundation to analyze the six known, planetary orbits. A summary of Kepler's principles include:

- planetary orbits are elliptical with the Sun at one, common focus,

- a satellite sweeps out a constant area in equal intervals of time, and

- a ratio of the cube of the average, orbital separation for a satellite and the square of its orbital period is a constant value.

During Kepler's search and following the identification of his three principles of orbital motion, Kepler continued looking for a more inclusive,geometric connection. Authors[1] writing about Kepler's life and works note this search for an invisible structure, a harmony among the spheres.

The ellipse is an awesome piece of geometry. Its 'top-bottom' and 'left-right' symmetry is compelling and quite beautiful. If one is to design a new model, a quantum gravitation model, the beginning must include a return to Keplerian geometry, searching for a logical variation upon which to form a new hypothesis. Kepler's first principle, **planetary orbits are elliptical with the Sun at one, common focus**, provides insight to frame a question. This expression states a purpose or function for one focus; but, there is no physical significance associated with the second focus. Therefore, Kepler's first principle of orbital motion is an asymmetric statement about a symmetric geometry. What is at the other focus? Astronomers and others probing the sky with telescopes have not found any physical component associated with this second focus. Could some invisible structure be located and/or associated with this focus?

A resonant system, a harmonic system, and a quantum system are all systems having a characteristic periodic motion or a measurable frequency. Could the structure and dynamics of the Solar System be an example of a quantum, gravitational system? Eintein and others[2] have hypothesized that gravitational waves are likely to be associated with binary stars. Jupiter is to small to form a binary with the Sun, but its mass is sufficiently large to qualify for a substar status.

The Solar System is not a binary system, but with the Sun-Jupiter coupling comes the possibility of gravitational waves within its structure. So begins a process of forming an inquiry model: framing a question, forming assumptions, and organizing operational hypotheses to orchestrate a search for evidence of gravitational waves within the Solar System.

To form a gravitational wave model new assumptions and hypotheses describing an orbital system must be proposed. Within the Solar System there must be a modification for Keplerian, orbital geometry. Connections between wave characteristics and a new geometry must yield results that are consistent with Kepler's original principles, new predictions, and new experimental results further describing structure and/or function within the system.

Designing a feasibility study to test assumptions and clarify hypotheses is a difficult task. This level of trial-and-error activity includes: 1) designing and building appropriate equipment to simulate the inferred geometry, 2) studying and analyzing the simulated, wave effects, and 3) making adjustments in the physical equipment, the intellectual probing, and an identification of possible connections and interrelationships. A wavetank, fitted with the following ancillary components, was selected to study wave interference patterns and the hypothesized geometry:

- an elliptical boundary (a rubber pressure-cooker seal) defining a path of destructive interference, and

- a pulsating energy source at one focus.

A Feasibility Study: Wave Interference within an Elliptical Boundary

Using a standard ripple tank, a rubber, pressure cooker seal for the forming of an elliptical barrier, and a single, periodic point source set to pulse at the focus of the ellipse, the investigation of wave interference within an elliptical structure began. To document the interference patterns and ancillary effects photographic and video filming is recommended. Following is a summary of the identifications made during the feasibility study:

- waves move radially away from the source, reflecting from the elliptical interface (nodes), and move toward the second focus,

- water surface at the two foci is pulsing up-and-down, moving out-of-phase for different, ellipse eccentricities (**note:** cover photograph), and

- interference patterns form hyperbola and parabola-like-structures providing evidence for paths of minimal, potential energy (hyperbolae) and structures similar to electromagnetic lines-of-force (parabolas).

All models have components with analogous, 'real world' counter-parts. Astronomers and/or cosmologists design their models to describe an effect or process within the macrocosm. This feasibility study was designed to gather information about the behavior of waves within an elliptical boundary and provide evidence to form a wave model to describe gravitational interactions in the Solar System. Components in the feasibility study were hypothesized to have the following cosmological counter-parts: 1) the water surface and the space-time continuum, 2) a pulsing, energy source and the Sun (a source of gravitational waves), and 3) the elliptical barrier and a planetary orbit (a path of low, gravitational potential energy).

Albert Einstein' s relativity theory predicted cosmological structures affect the transfer path for the propagation of energy between points in space. The collection of possible energy paths from a source forms a pliant, space-time manifold or continuum. Einstein predicted large objects: black holes, stars, planets, and moons produce a negative distortion (a depression) on the space-time manifold. In 1916 Einstein used logic from his relativity theory to calculate an expected 1.74 arc-second distortion in the space-time continuum due to the Sun's mass. In 1919 Sir Athur Eddinton organized expeditions to South America and Africa to photograph the solar eclipse, looking for optical evidence to confirm Einstein's prediction. The photographs provided optical evidence for slight, locational shifts in the star field surrounding the Sun, confirming Einstein's prediction.

The inquiry proceeds with a formulation of a set of assumptions (**if** statements) and a hypothesis (a **then** statement), **if:**

• the Sun's location on the space-time manifold is a depression,

• the space-time continuum within the Solar System is nearly flat, and

• perfect symmetry is a mandate for an elliptical structure,

then the second focus associated with each planetary orbit should be a positive distortion on the space-time manifold. Therefore, a mass element passing close to each of these locations (second focus for each orbit) should be deflected away from the location.

Planetary Orbit: An Ellipse with the Sun at One, Common Focus

Kepler had intuitive thoughts about the planets experiencing some form of resonant event. He likely thought about a special relationship between the sun and each respective planet and among those planets known to have one or more satellites. Do the orbital characteristics have any other connections with the other curves forming the conic sections: the parabola and/or the hyperbola? By the end of Kepler's life in 1630, Galileo Galilei was identifying and analyzing a parabola as a best-fit curve to describe the trajectory for a mass transfer near the Earth's surface.

Since the early 1600's many have pondered the possibilities of a 'harmony among the spheres.' While additional information has been collected about planetary orbits, an identification of orbital resonance remains only conjecture and speculation. A forthcoming gravitational wave model for the Solar System could provide guidance to search for evidence of gravitational waves. Such experimental evidence would provide physicists with a window to observe and characterize gravitational interactions. This could be an important step for science and its quest to unify the four forces used to describe natural cause-and-effect.

Intellectual inferences have begun offering an order-of-magnitude value for wave characteristics and/or resonant effects[2] associated with gravitational waves. But, the evidence is sparse and too inaccurate to design instrumentation to measure gravitational wave characteristics. The electromagnetic field theory sprung forward when instrumentation to synthesize and measure very small wave manifestations was designed and tested. But, to design, build, test, and modify equipment to measure large, scale quantities like those associated with the waves of gravitation becomes a most difficult task. Current technology is stretched to provide accurate, separation measurements for the planets and the sun. We can only approximate this separation by secondary methods employing mathematical techniques, calculations from a mathematical model. Small percentages in error, even to six orders of magnitude less than one, while acceptably small for most laboratory investigations of electromagnetic phenomena, yield a rather large variance in the cosmos. Making intellectual adjustments to accommodate for these differences must be tempered by looking for common trends and/or patterns in the results associated with gravitational interactions.

Properties and characteristics associated with an ellipse

The geometry of the ellipse is standard in most mathematics and physics courses. Any wave hypothesis for gravitation will be forced to manifest descriptions consistent with Kepler's orbital principles and the geometry of an ellipse. Following is a diagram conveying common characteristics associated with an ellipse:

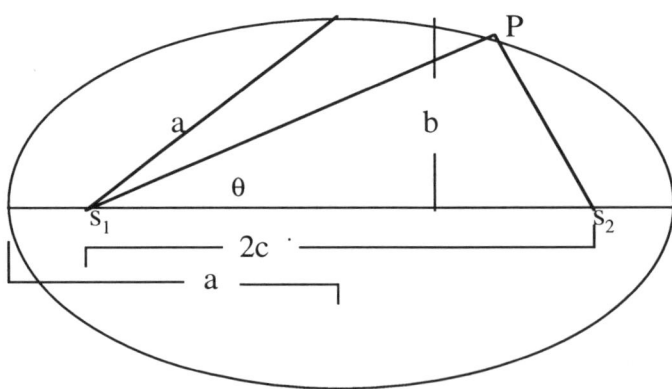

a -> semi-major axis length   b -> semi-minor axis length

s_1 -> primary focus (sun) s_2 -> secondary focus (?)

2c-> foci separation,   P-> orbital point, and θ-> angle between PS_1 and 2a.

Figure 6.1: General properties of an ellipse.

Wave interference models for two point source interference[3] form interference patterns in the shape of hyperbolae. This model will propose a two point source model with the sources out-of-phase, having a phase delay of one-half, $p = 1/2$. Welding this assumption into the geometry of the ellipse advances the emerging model toward a mathematical synthesis, a relationship formed through a connection of an ellipse and a hyperbola.

A Parametric Relationship: An Ellipse - Hyperbola Intercept

- about an ellipse

  1) $PS_1 + PS_2 = 2a$ with $(0,0)$ at the ellipse center, and

  2) ellipse (orbital) eccentricity, $e = c / a$.

- about a hyperbola

  1) $PS_1 - PS_2 = 2A$ with $(0,0)$ at the ellipse center, and

  2) hyperbola eccentricity $= c / A$.

- parametric solution for the intercept of an ellipse and a hyperbola

  1) $2 PS_1 = 2 (a + A)$ or $PS_1 = a + A$,

  2) $PS_2 = a - A$, and

  3) $A = a [[(1 - e^2) / (1 - e \cos \theta)] - 1]$.   (**note: Appendix: A2.1**)

An Argument for Reflection at an Elliptical Interface

If energy moves outward from the point source s_1, a straight-line path from s_1 to any point on the ellipse and then on to point s_2, forms an angle, $s_1 P s_2$. The normal to the tangent line drawn through the point P on the ellipse bisects the angle, $s_1 P s_2$, forming the incident and reflected angles for a particle transfer along the path $s_1 P s_2$ (**note: Appendix A 2.3**). Results from the feasibility study infer this wave form to be perfectly out of phase with the incident energy coming from s_1.

Following is a diagram demonstrating the geometry for energy moving from a focus (s_1) to an elliptical interface and reflecting to the second focal point. The energy propagating from s_1 interferes with the energy coming along the reflective path, from P to s_2. The pathlength difference ($\Delta l = | Ps_1 - Ps_2 |$) produces a phase shift of $1/2$; the pulsating energy emitted from source s_1 is out-of-phase with the energy arrival at s_2.

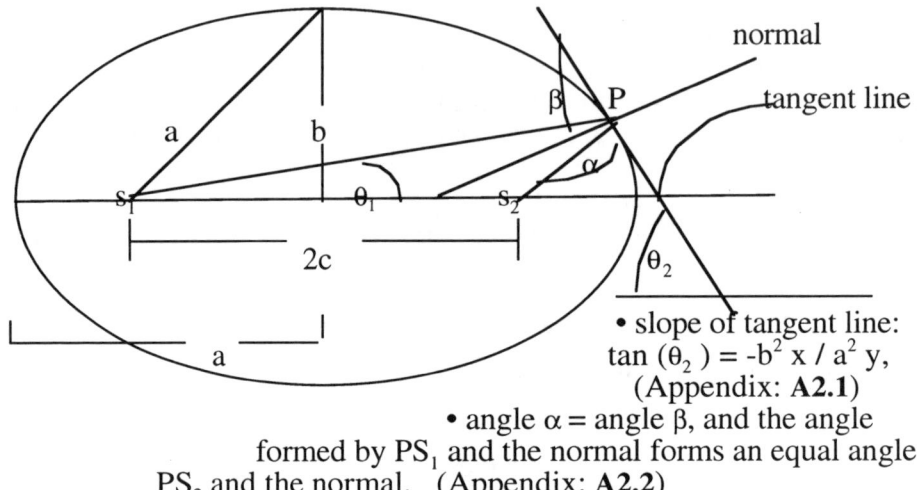

slope of tangent line:
$\tan(\theta_2) = -b^2 x / a^2 y$,
(Appendix: **A2.1**)
• angle α = angle β, and the angle
formed by PS_1 and the normal forms an equal angle
PS_2 and the normal.   (Appendix: **A2.2**)

Figure 6.2: Tangency characteristics for a point on an ellipse.

Designing A Computer Model to Search for Predicted Outcomes

Good models demand a connection with reality. One such linkage comes through a prediction from a synthesized model and a compatible pattern and/or trend found by searching through or among existing data. Two, point sources oscillating in a wave tank produce nodal paths similar in shape to hyperbolae; and, one point source oscillating at one focus within an elliptical boundary also generates paths of interference similar in structure to both hyperbolae and parabolas (note: diagram on the front cover). So, a simultaneous solution for the intersection of a hyperbola and the ellipse is a valid step for the analysis.

A number of computer programs (mathematical models) were written to test the wave, interference model. The database (**Appendix: A1.1**) for each program is information about the Solar System commonly found in most astronomy textbooks. Each program was designed to search for patterns among the data, looking for consistent patterns and/or fundamental quantities to associate with a wave structure or a wave view for gravitational events within the Solar System. Following is a listing of the primary, geometric connections:

an ellipse --> $PS_1 + PS_2 = 2$ a,

a hyperbola --> $PS_1 - PS_2 = 2$ A, and an interference pattern and a

hyperbola --> $PS_1 - PS_2 = 2$ A = n λ (a pathlength difference). A geometric model and/or mathematical model incorporating hypothetical connections can be summarized as follows:

- hyperbolae, interference nodal lines, form between the two foci and govern their intersection with the elliptical, orbital path,

- the foci are out-of-phase, suggesting an incorporation of odd-integer interval steps between the foci (separation 2c), and

- the parameter 'A' for a given hyperbola is the separation of the vertex and the center of the ellipse along major axis. This parameter is the path-length difference for a given interference pattern and is the basis for the assumption '2A = n λ' in the computer model. The simultaneous solution for the equation of the ellipse and the equation of the hyperbola provide another mathematical connection to be included in of the design of the computer programs.

$$PS_1 + PS_2 = 2\,a\,,$$

$$PS_1 - PS_2 = 2\,A = n\,\lambda\,,\quad\text{and}$$

$$2|PS_1 - a| = 2c = n\,\lambda\,.$$

Since the symmetry for an ellipse is an important consideration, the integer factor 'n' becomes an odd-integer multiple (1, 3, 5, etc.). Using the '2c' foci separation and the pathlength difference information, n λ = 2A, a computer designed to search for a largest, common factor among the data began. Also, remember the common structural form for the electrostatic force and the gravitational force, the particle view of an electron orbital used by Niels Bohr to describe the structure of a hydrogen atom was given significance. These include:

- the electron-proton interactions may result from electromagnetic, wave interference, likely constructive interference,

- the orbital circumference for the electron is an integer multiple of the electron wavelength,

- the orbital period for the electron at alternative separations also forms a constant R^3/T^2,

- the upper limit for electron speed is the ground state structure is a speed of 2.1878E6 m/s, and

- electron speeds greater than the speed associated with the Bohr orbital speed are associated with the electron as it spirals toward the proton and a transformation into a different component of matter, a neutron by the 'K-capture' process.

Trial, analysis of data output, program modification, and new trial can be a long, arduous cycle. In time, data trends and patterns began to suggest the inclusion of filters within the program to make the search more efficient. These filters become analogous to boundary conditions for the operation of the mathematical model. The design of these filters incorporated some of the arguments presented in the Bohr model. This type of controlled search within the structure of a computer program provides a significant bias for the output data and interpretation. But, that is one of the more interesting aspects of scientific modeling and mathematical modeling in particular.

Following is a diagram and summary of the assumptions associated with the design of the computer programs.

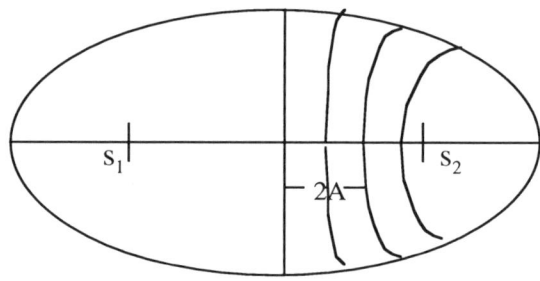

- points along the orbital path are points of destructive interference; points of low, gravitational potential energy,

- an odd number of integer wavelengths between the foci, s_1 and s_2,

- destructive interference model with $PS_1 - PS_2 = 2 A = n \lambda$, and

- hyperbolae are nodal lines of constant gravitational potential energy.

With the data filters and supportive assumptions in place, a comparison began, a search for a best-fit connection among the data and the guidelines used to describe the wave, interference model. The output of results from the computer search among the Solar System data presented values forming an interesting and recurrent pattern. Following is a summary from one computer search:

Planet	'2c' Interval (m)	Integer	Separation (m)
Venus	1.4717E9	17	8.6571 E7
Earth	4.9966E9	57	8.7660 E7
Mercury	2.3813E10	273	8.7225 E7
Mars	4.2579E10	485	8.7790 E7
Neptune	7.3871E10	841	8.7835 E7
Jupiter	7.5339E10	859	8.7705 E7
Saturn	1.5523E11	1767	8.7850 E7
Uranus	2.6450E11	3011	8.7845 E7
Pluto	2.9278E12	33323	8.7861 E7

Table 6.1: The summary of a search for intervals between orbit foci for each planet in the Solar System.

The separations listed in Table 6.1 could be analogous to the separation of nodes in a resonant, interference pattern. Twice this separation would be the fundamental wavelength for the harmonic system. This value could also be associated with the fundamental wavelength for a gravitational standing wave within the Solar System. This wavelength is analogous to the circumference of the fundamental orbital in the Bohr model for a hydrogen atom. Using the orbital speed of 2.1878E6 m/s as the limiting speed for each fundamental, planetary orbital, the following averages can be defined from the table of values:

- $\lambda_1 = 1.7519E8$ m,
- $r_1 = 2.7882E7$ m,
- $T_1 = 80.26$ sec, and
- $v_1 = 2.1828E6$ m/s.

Summary and Speculation

Constructing a model in uncharted waters is risky business. Niels Bohr began his model for the hydrogen atom with such trepidation. Being wrong within a professional community carries a sigma, but the fragrance of success is a powerful motivation. Many successes in science, and elsewhere, have risen from the ashes of failure and defeat. A wave model for gravitational interactions would be very helpful to physics and cosmology. But, like Bohr's introductory model for elemental hydrogen, a wave model for gravitation will likely receive notable attention and severe criticism. With regard to any on going process like modeling, it is important to remember: An ending only follows a beginning.

For an emerging model the gates of assumption are tethered to pliant poles, not to pillars of certainty. Progress requires a refiner's fire, additional work to modify and clarify each part of the model. If the assumptions are reasonable, then any calculations should begin to illuminate a good result. A model is like Swiss cheese, enough substance to be encouraged; yet, an awareness of gaps in logic and reason for concern. The process begs for more information: more data, better data, additional experimentation, more development, particularly theoretical development. At each step of this progression is a call for identification, definition, and/or connection with earlier, experimental results.

This wave model is a corollary to Kepler's view of planetary orbitals and offers a modified geometry to accommodate a wave view for gravitational interactions within the Solar System. The modifications to Keplerian geometry include:

- the two foci within a planetary ellipse are out of phase structures,

- all points along the orbital path are points of destructive interference; low, gravitational potential energy,

- an odd number of integer wavelengths between the foci, s_1 and s_2,

- destructive interference model with $PS_1 - PS_2 = 2 A = n \lambda$, and

- hyperbolae are nodal lines of constant gravitational potential energy.

An interesting aspect associated with the preceding result suggests an orbital separation well within the photosphere of the Sun. The orbital separation value, $r_1 = 2.7882E7$ m, is approximately **4.0%** of the radius of the Sun's photosphere. Any significance for this value and its coupled response within the Sun is unknown. But, this value is about mid-range for the minima reported in the 'center-of-mass' study for the Solar System (**note:** Graph 4.3, page 55). One inference from the trend of data associated with that study could be a transitional state existing at about 0.02 solar radii, $r \approx 1.3939E7m$. Perhaps this transition is associated with a change from electromagnetic interactions to nuclear interactions within the Sun.

One of the compelling interests within astrophysics is the structure and function of the solar interior. Helioseismology is one of the newer and more powerful tools for inquiry used by these scientists. The results are encouraging but most data is laced with minor variations (weak signals) and high uncertainty. The influence of the orbiting planets on the operation of the Sun is hypothetical. Yet, most scientists believe cause-and-effect science is as real within the Sun as it is within their home. The search for hidden, solar effects continues with confidence and hope; solar composition and structure are slowing yielding to description and interpretation.

Lastly, comes the projected wavelength for a resonant, standing wave within the Solar System, $\lambda_1 = 1.752E8m$. Scientific thought and experimentation must explore the field of gravitation in greater detail, searching for additional evidence to confirm or discount this result. Chapter Seven will use existing data for the Solar System to enlarge an argument for this quantity and for a quantum view of gravitation.

Resources and References for Chapter Six:

1. Koestler, A., <u>The Watershed</u>, (Lanham, Maryland: University Press of America, Inc, 1960), p 202-226.

2. Misner, C., Thorne, K., and Wheeler, J., <u>Gravitation</u>, (San Francisco, California: W.H. Freeman and Company, 1973), p 986.

3. Haber-Schaim, U., Dodge, J., & Walters, J., <u>PSSC Physics</u> (6th ed.), (Lexington, Massachusetts: D.C. Heath and Company, 1986), p 492.

Chapter 7

A QUANTUM, GRAVITATIONAL MODEL FOR SATELLITE MOTION WITHIN THE SOLAR SYSTEM

Introduction

> **"The heavens declare the glory of GOD; the skies proclaim the work of His hands. Day after day they pour forth speech; night after night they display knowledge. There is no speech or language where their voice is not heard. Their voice goes out into all the earth, their words to the ends of the world." (Psalm 19:1-4, NIV)**

The motion of the planets about the Sun, a particle view, was used by Niels Bohr to design his analog for electron, orbital motion in a hydrogen atom. This model uses the quantum view of the hydrogen atom as an analog to present a quantum, gravitational view for the planets in orbit about the Sun. The model incorporates new, geometric arguments to embellish the Keplerian geometry for planetary orbits, principally a resonant, gravitational wave interference strategy to account for planetary motion in the Solar System. This model draws from classical physics and incorporates basic principles of quantum physics to introduce a quantum, gravitational view for the space-time continuum, satellite orbital motion, and structure within the Solar System.

This study is not focused on the presentation of an end result. Rather, its purpose is to establish a beginning for an inquiry model for a quantum, gravitational view of the Solar System. Chapters five and six provide a platform for the formulation of this model; a resonant, gravitational wave, interference model with assumptions and operational guidelines. The model becomes a platform to design computer programs with an inclusive purpose: 1) to incorporate this hypothetical geometry among the known planetary and satellite orbits and 2) search for commonalties among the parameters used to describe planetary orbits looking for evidence of gravitational standing waves within the Solar System. The search compares and analyses planetary, orbital characteristics including:

- the foci separation $(2c/n)$,
- twice the orbital area per unit time $(2A/T)$,
- incremental, orbital mass (M_p/n),
- orbital speed $(n\,v_p)$,
- orbital period (T_p/n^3), and
- orbital perihelion (r_p/n^2).

In addition to analyzing the orbital motion of the planets in the Solar System, the model was used to analyze the orbital characteristics associated with the moons of Jupiter and predict some of the gravitational characteristics for the Jovian System.

A circle is a special form of ellipse. It forms when the two foci are super-imposed one on top of the other; making its eccentricity zero. If the eccentricity of an ellipse is greater than zero and less then one, the two foci have a finite separation. This gravitational wave model incorporates Kepler's view of planetary orbits and incorporates the following two arguments:

- gravitational potential energy propagates inward toward the Sun and toward the central body for each of the respective satellite systems, and

- the source of gravitational lines of force is associated with the second foci for each satellite orbital.

Therefore, an analysis of the space between these two foci is a good choice to begin this analysis. This result was summarized in tabular form in **Table 6.1** and is modified here to report a predicted value for the wavelength of a gravitational standing wave. A resonant gravitational wavelength was calculated using the following algorithm:

$$\lambda = 2 \, (2c \, / \, n) = 4c \, / \, n$$

An Analysis of Orbital Foci Separation

Planet	'2c' Separation (m)	Integer	Wavelength (m)
Venus	1.4717E9	17	1.7314E8
Earth	4.9966E9	57	1.7532E8
Mercury	2.3813E10	273	1.7445E8
Mars	4.2579E10	485	1.7558E8
Neptune	7.3871E10	841	1.7567E8
Jupiter	7.5339E10	859	1.7541E8
Saturn	1.5523E11	1767	1.7570E8
Uranus	2.6450E11	3011	1.7569E8
Pluto	2.9278E12	33323	1.7572E8

Table 7.1: An analysis using the computer model for foci separation (2c / n) with an error window of 1.5%. Averages for these values are based on an integer search between the foci for a best-fit wavelength and calculated, Keplerian values associated with each planetary orbit:

- $\lambda_1 = 1.7519E8$ m,   - $r_1 = 2.7882E7$ m,

• T_1 = 80.26 sec, and   • v_1 = **2.1828E6 m/s**.

The selection of the value for 'v_1 = 2.1826E6 m/s' and its association with the speed of light and the fine structure constant, $v_1 \cong c/137.036$, becomes an instant focus. As will be pointed out in subsequent steps, this speed became a recurrent speed among the output values associated with a resonant, wave model; hence, the computer model used to calculate and search for a best-fit value among the orbital parameters associated with each plane. It infers a connection for the physics describing orbital motion in the microcosm and the macrocosm.

Satellites move about a central body in elliptical orbits, sweeping out equal areas in equal intervals of time.

This is an astounding piece of geometric analysis and it requires the central mass to be approximately one-thousand times more massive then the mass of the coupled satellite. Within the microcosm the particle view for the proton-electron coupling has a mass ratio of approximately 1836, $m_p / m_e \cong 1836$. The ratio of the mass of the Sun to Jupiter is near 1040. Perhaps this is a characteristic for all orbital systems. The orbital is dependent on the defining characteristics of the central body in relationship with its associative satellite, whether within the influence of an electromagnetic force in the microcosm, or a gravitational effect within the Solar System and/or the macrocosm.

Classical physics proposes a common mathematical form for the forces orchestrating the orbital motion in the Bohr model for a hydrogen atom and a planet moving about the Sun. Both forces have an inverse square, radial separation controlling the point-source geometry surrounding the central mass. Therefore, orbital characteristics within these two realms should exhibit analogous geometric and structural symmetries and/or effects. These analogous, orbial characteristics should present a pathway to link these vastly different, structural systems.

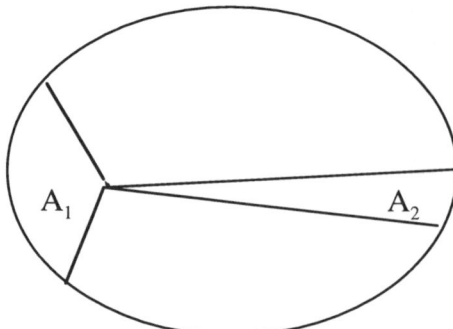

Figure 7.1: A representation of equal areas swept out by an orbital element in equal intervals of time.

Remember, the area of an ellipse is dependent on the ellipse lengths of each the semi-major (**a**) and semi-minor (**b**) axes. The area inscribed within an ellipse is given by approximation to be $A = \pi\, a\, b$.

The concept of area-per-unit-time (A/T) is fundamental to Johannes Kepler's second principle of orbital motion. While a complete interpretation for Kepler's three principles of orbital motion may not exist, the following relation-ships are commonly associated with Kepler's second principle:

- $dA / dt = 1/2\ (r\ v)_p = 1/2\ (r\ v)_a$ (at perihelion or aphelion),

- $dA / dt = L / 2m$ (1/2 angular momentum / mass), or

- $2\ [dA / dt] = L / m = (r\ v)_p = (r\ v)_a$ (angular momentum / mass).

These relationships mask an affect for mass. The first statement does not have a mass component and the second and third have a mass factor (mass of satellite) which occurs in both the numerator and denominator of the mathematical expression, producing the factor one and inferring a trivial or non-existent contribution or effect for the orbital mass. Therefore, Kepler's work does not suggest a dependency on the mass of the orbital element within his description of orbital motion. Meanwhile, Isaac Newton's synthesis of a gravitational model incorporated an effect for coupled masses in a system, yielding his historic relationship:

$$F_g = k_1\ m_1\ m_2 / r^2.$$

Within an orbital system, m_1 and m_2 are constants and can be united to form a single constant; therefore, making the gravitational force proportional to the inverse square of the orbital separation:

$$F_g = k_2\ [1/ r^2] = G\ M_1\ m_2 / r^2$$

This logic paved the way for Newton to calculate a value for the universal grav-itation constant, a value for the mass of the Sun, and a mass for each planet. Newton's model presented a 'cause' for the 'effect' associated with orbital mo-tion and provided an insight to link Kepler's work with Newton's mathematical model for gravitation. The following algorithm demonstrates this linkage:

$$K = R^3 / T^2 = G\ M_1 / 4\ \pi^2$$

An analysis of the 'area/time' for each orbit provides evidence for a possible quantum effect within a gravitation system. Therefore, the computer search pro-posed within this model includes the following arguments:

1) the area/time for an ellipse, $dA/dt = \pi\ ab / T$, and

2) the orbital angular momentum per unit mass and the quantity 2A/T.

The Bohr model identifies a quantized value for angular momentum; therefore, an analogous search for a quantized value among the 2A/T, orbital values in the Solar System yields the following information:

1) orbital speed --> v_1 = 2.1878E6 m/s   (**note:** about 1/137 light speed),

2) orbital radius --> r_1 = 2.774E7 m   (**note:** within the sun),

3) orbital period --> T_1 = 79.83 sec, and

4) wavelength ---> λ_1 = 1.746E8 m.

The argument also assumes a resonant wavelength for the gravitational wavelength associated with the central mass being equivalent to the circumference of a fundamental orbital. The radius of this orbit is about 4% the radius of the Sun's photosphere and may mark a boundary for an electromagnetic influence.

The output from each of the computer programs inferred this pattern and facilitated the analogous connection to Bohr's model. Quantum electrodynamics repeatedly displays the fine-structure constant within its mathematical descriptions for structure and dynamics in the microcosm. The symmetry of argument inferred a search for the following fundamental quantities associated with planetary orbits in the Solar System:

- the foci separation, $2c / n$,
- orbital mass, M_p / n,
- orbital speed, $n\, v_p$,
- orbital area per unit time, $2A / T$,
- perihelion, r_p / n^2, and
- orbital period, T_p / n^3.

> **Analysis: Planetary Area per unit Time**

Planet	Integer	2A / T (m² / s)
Mercury	45	6.0277E13
Venus	63	6.0151E13
Earth	74	6.0206E13
Mars	91	6.0175E13
Jupiter	168	6.0451E13
Saturn	228	6.0478E13
Uranus	323	6.0584E13
Neptune	404	6.0676E13
Pluto	446	6.0626E13

Table 7.2: Analysis using the computer model for orbital area / period, (**2A/T / n**), with an 'error window' of 1.5%. Averages for this search and comparison include:

- $2A/t = 6.04E13 \ m^2 / s,$
- $r_1 = 2.78E7 \ m,$
- $\lambda_1 = 1.75E8 \ m,$
- $T_1 = 80.11 \ sec,$ and

$$v_1 = 2.1839E6 \ m/s.$$

The inference of a limiting orbital speed for a mass element in a fundamen-tal orbital in both the microcosm and macrocosm added significance to the sym-metry of the force relationships indicated in **Figure 5.2**. Also, it permitted con-sideration of a connection for a resonant force controlling planetary, orbital configuration. The formation of a gravitational model that was analogous with the Bohr model was advancing in a reasonable direction.

One notable connection passes through the concept and value associated with the 'quantum of circulation, $h_{bar} / m_0 = 1.1574E-4 \ m^2$ /s. By analogy there should be a correspondent value for the Solar System. The 2A/t value sum-marized in the preceding table should be related to this value:

$$2 \ A \ / \ T \ = i_{bar} \ / \ m_0 \ = \ 6.04E13 \ m^2 \ / \ s = v_1 \ r_1,$$

where i_{bar} is a fundamental value for orbital angular momentum in the Solar System and m_0 is a fundamental limit for a smallest, stable orbital mass associated with the primary mass, the Sun.

> **Minimum, Stable Orbital Mass About the Sun**

Planet	Integer	Mass (kg)
Mercury	309 *	1.0294E21
Mars	623	1.0302E21
Pluto	1048	1.0305E21
Venus	4735	1.0313E21
Earth	5798	1.0312E21
Uranus	84189	1.0313E21
Neptune	99587	1.0313E21
Saturn	551171	1.0313E21
Jupiter	1843377	1.0313E21

Table 7.3: An analysis using the computer model for planetary mass (M_p / n), with an error window of 0.05%. The average minimum, increment for orbital mass within the Solar System is $m_1 \cong 1.03E21$ kilograms.

[**note:** * value adjusted from 310 for a 'best-fit' with the trend of data from the computer model]

Table 7.3 is a summary of a computer search for information about size, stability, and possible limits for planetary mass in the Solar System. This value could be a minimal mass in a stable orbit and/or a maximum mass associated with some internal, dynamic effect within the Sun. A hypothesis for this effect follows; and, it may provide an insight into the quantization of orbital, angular momentum within this system.

A technique used to modify the integer value for entry into the computer model, n = '310' for Mercury, actually involved charting output values on logarithmic, graph paper. Charting several planets generated lines for each planet appearing to be straight and parallel. Using the slopes of these lines and the location of the quantum orbital two, a parallel, construction line gave an intercept of about 2.0E21 kg. Since Mercury is the smallest planet, the value in question involved this planet and its integer intercept value of about '310.' Therefore, the graphical analysis suggested a prediction of about 1.0E21 kg and a prediction for the size of a fundamental, orbital, mass element prior to the computer search trial.

The computer output value also seemed to form a near limit with the integer value of 310 (actually, a best integer value of 309) and a mass of 1.03E21 kg. This value, a smallest increment of orbital mass within the Solar System, provided a value for substitution in the algorithm for a minimal, orbital angular momentum:

- $L_1 = (2A/T) (m_1) = v_1 r_1 m_1$ or   momentum,
- $i_{bar} = L_1 = (6.04E13 \ m^2 / s) (1.030E21 \ kg) = 6.22E34 \ kg \ m^2 / s.$

As was mentioned earlier, this model is very rough and its primary intent is to provide a pathway for physicists to press the inquiry toward truth and understanding. To give an interpretation for these, Solar System values, a fundamental value for stable, orbital mass and an increment of orbital angular momentum, is reaching deep into the pocket of hypothesis, if not a hoped outcome. Since these values are located within the Sun, perhaps they contribute to and/or effect the internal structure and operation of the Sun.

From the results reported in Chapter Four, it seems probable the Solar System's center-of-mass moves in a spiral, looping, pattern in-and-out of the Sun's photosphere. The periodicity from a major 'maximum to maximum' displacement is approximately eighty years, about seven Jupiter periods.

Scientists are confident defined zones exist within the Sun. These zones are difficult to define, locate, and interpret. But, it is not beyond the realm of possibility for planetary, orbital variations to have an impact on the internal dynamics and/or operation of the Sun. Should this linkage become more evident in time, these values may be modified or take on added significance.

These minimal values could be involved in a 'quantum of circulation' with-in a gravitational system. Planetary orbits, could be described or characterized with these values, quantized, gravitational values for the Solar System. Since angular momentum has parameters associated with its orbital nature, its value may incorporate sets of quantized values. From the logic summarized in Chapter 3, comes a suggestion of the following mathematical structure for orbital, angular momentum in the Solar System:

$$L_0 = n_1 \, n_2 \; 6.22E34 \text{ kg m}^2 / \text{s} \qquad \text{where,}$$

L_0 --> quantum of orbital angular momentum,

n_1 --> an integer associated with mass, and

n_2 --> an integer associated with orbital geometry in the Solar System.

The final three tables of values display evidence for a quantum, gravitational structure in the Solar System. The results link together nicely and provide a reasonable way to summarize all of this data and perhaps the mathematical models used in the computer search for fundamental values associated with satellite orbits in the Solar System.

> An Analysis of Orbital Speed

Planet	Integer	Orb speed (m/s)
Mercury	45	2.1993E6
Venus	62	2.1711E6
Earth	73	2.1736E6
Mars	90	2.1804E6
Jupiter	166	2.1692E6
Saturn	225	2.1719E6
Uranus	319	2.1702E6
Neptune	400	2.1711E6
Pluto	444	2.1709E6

Table 7.4: An analysis using the computer model for planetary speed ($n \, v_p$), with an 'error window' of 1.5%. Averages for the search and comparison of an incremental speed for orbital motion include:

- $\lambda_1 = 1.7640E8$ m,   • $r_1 = 2.8075E7$ m,

- $T_1 = 81.10$ sec, and   • $\mathbf{v_1 = 2.1753E6}$ m/s.

Again we see the potential for an association with planetary orbital speed and a fundamental, orbital speed close to the 2.1878E6 m/s. This speed along with its associative radial separation may represent a boundary condition marking the edge of an interface between the nuclear region and the electromagnetic region. Matter having speeds less than the Bohr, orbital speed are within the region dominated by electromagnet interactions. Should the particle speed increase beyond this resonant, threshold value, the particle would spiral into a region of nuclear interactions.

An example model for this type of interaction is 'K-capture' in the beta-decay model. Conjecture for a gravitational analog comparing with 'K-capture' with a gravitational interaction could include the following components: a driving force associated with a central mass, and an orbital element interacting with the primary and moving into the realm of the nuclear force.

> **An Analysis of Orbital Period**

Planet	Integer	Orb Period (s)
Mercury	46	78.1041
Venus	63	77.6418
Earth	74	77.8783
Mars	91	78.7654
Jupiter	168	78.9433
Saturn	227	79.4731
Uranus	322	79.4081
Neptune	403	79.4531
Pluto	462	79.4773

Table 7.5: An analysis using the computer model for planetary period (T_p / n^3), with an 'error window' of 1.5%. Averages for the orbital periodicity include:

- $\lambda_1 = 1.73E8$ m,

- $r_1 = 2.75E7$ m,

- $T_1 = 78.79$ sec, and

- $v_1 = 2.19E6$ m/s.

> An Analysis of Perihelion Separation

Planet	Integer	Perihelion (m)
Mercury	41	2.7388E7
Venus	62	2.7932E7
Earth	73	2.7598E7
Mars	86	2.7929E7
Jupiter	163	2.7870E7
Saturn	220	2.7882E7
Uranus	313	2.7951E7
Pluto	398	2.8049E7
Neptune	399	2.8034E7

Table 7.6: Analysis of a Computer Model for Planetary, Perihelion Separation (r_p / n^2), with an 'error window' of 2.5%. Averages for the search and comparison of orbital perihelion include:

- λ_1 = 1.75E8 m,
- r_1 = 2.78E7 m,
- T_1 = 80.11 sec, and
- v_1 = **2.18E6 m/s**.

Notice the single step difference between the iteration associated with the perihelion separation for Pluto and Neptune. This is a very interesting result. It may be associated with Pluto's apparent, orbital overlap with the orbit of Neptune.

Summary of Orbital Parameters within the Solar System

Model Analysis:	Wavelength (m)	Orbital Radius (m)	Orbital Period (s)	Orbital Speed (m/s)
Wavelength	1.7519E8	2.7882E7	80.26	2.1828E6
2 A/T	1.7495E8	2.7844E7	80.11	2.1839E6
Orb Speed	1.7640E8	2.8075E7	81.10	2.1753E6
Orb Period	1.7304E8	2.7541E7	78.79	2.1962E6
Perihelion	1.7497E8	2.7848E7	80.11	2.1841E6

Table 7.7: A summary of the analyses using the computer model to search for fundamental, planetary orbital characteristics in the Solar System.

The Universal Gravitation Constant

In 1665 Newton's model for gravitation hypothesized a value for the universal gravitation constant. Lord Cavendish arrived at an experimental result a century plus later in 1798. One of the more common mathematical relationships to calculate a value for 'G' is as follows:

$$Force_{centripetal} \ = \ Force_{gravitational}$$

$$m \, v^2 / r \ = \ G \, M_s \, m_p / r^2 \qquad and,$$

$$G \ = \ r \, v^2 / M_s$$

The relationship between **r** and **v** is rather simple if the orbits are assumed to approximate a circular shape. This assumption produces another variation for the gravitation constant **G**,

$$G \ = \ r \, v^2 / M_s = (4 \, \pi^2 \, r^3 / T^2) / M_s.$$

Easily identified in this expression is the factor, $r^3 / T^2 = 3.365E18 \ m^3 / s^2$, the Keplerian constant.

If a resonant, gravitational effect exists, then there must be another set of parameters whose mathematical grouping also must yield a common numeric value for the universal gravitation constant, $G = 6.672E-11 \ (m^3 / s^2) / kg$. Consider the following mathematical forms:

$G = (\lambda_1^2 r_1 / T^2) / M_s = (1.75E8m)^2 \, (2.7753E7m)/(80.11s)^2 \, / 1.991E30kg = 6.652E-11(m^3 / s^2)/kg,$

$G = r_1 v_1^2 / M_s = [(2.7753E7m)(2.1878E6m/s)^2]/1.991E30kg = 6.672E-11(m^3 / s^2)/kg,$

$G = (2A/T)(v_1)/M_s = [(6.041E13m^2 /s)(2.1878E6m/s)]/1.991E30kg = 6.638E-11(m^3 / s^2)/kg.$

Assuming the validity of this calculation, then in a gravitational resonant system the wavelength of a gravity wave must be proportional to the primary mass, M_p. The relationship becomes,

$$\lambda = 8.7543E-23 \ m / kg \ M_p.$$

With this result comes an option to calculate a resonant, gravitational wavelength for any mass.

> Predicted Wavelength for Standing Wave Associated with Each Planet

Planet	Mass[1] (kg)	Wavelength (m)
Mercury	3.1810E23	27.8
Venus	4.8830E24	427.5
Earth	5.9790E24	523.4
Mars	6.4180E23	56.2
Jupiter	1.9010E27	1.664E5
Saturn	5.6840E26	4.976E4
Uranus	8.6820E25	7.600E3
Neptune	1.0270E26	8.991E3
Pluto	1.0800E24	94.5

Table 7.8: Predicted wavelength for the gravitational standing wave associated with each planet in the Solar System.

Another important calculation is the gravitational wavelength associated with the Moon. Its mass is about 7.349E22 kg and according to the model its resonant, gravitational wavelength should be approximately 6.43 meters or about 21.10 feet. Identifying terrestrial manifestations associated with this feature pose an intriguing challenge and an important test for the validity of the model's logic. From a cursory view, cloud formations and glacial effects (particularly lakes and other features associated with large masses of ice) would be a likely locations to find possible multiples of this value.

Tidal forces which vary as an inverse cube of the separation between two bodies signal the significance of the moon in shaping surreptitious effects on or at the surface of Earth. This form of gravitational interaction may even impact structural limits for living systems.

> Speculation about 'The Roche Limit' for Jupiter[2]

Tidal forces are gravitational interactions between two large masses which may be form of a resonant effect among gravitational waves. Edouard Roche defined the calculation model for this effect in 1848:

$$L = [2.446 \; R] \; [D_p / D_s]^{1/3}, \quad \text{where the symbolic representations are}$$

- L - 'tidal extension' length,
- R - planet radius,
- D_p - planet density, and
- D_s - satellite density.

The Roche Limit for the planet Jupiter is approximately 1.7464E8 m. The calculated value for the standing wave using the computer, wave interference model is, $\lambda \approx 1.75E8$ m and this value is remarkably close to Jupiter's Roche Limit. Jupiter is the dominate planet within the Solar System; therefore, this apparent close agreement for two resonant, gravitational effects should not be a surprising result.

> A Proposed Investigation: The Ellipse, Saturn and an Empty Set

Any new model must measure up experimentally by demonstrating evidence to support its assumptions. The second foci associated with the orbit of Saturn should fall just outside earth's orbit about the Sun. Perhaps a probe could be designed to exit Earth orbit and explore this region of space. If this model is correct, one should expect a deflection of a small probe as it approaches this focus. The magnitude of the deflecting effect should be proportional to the mass of Saturn and may be larger due to the resonant nature of gravitational interactions within the Solar System.

Following is a crude diagram of this location with respect to Earth orbit[3]:

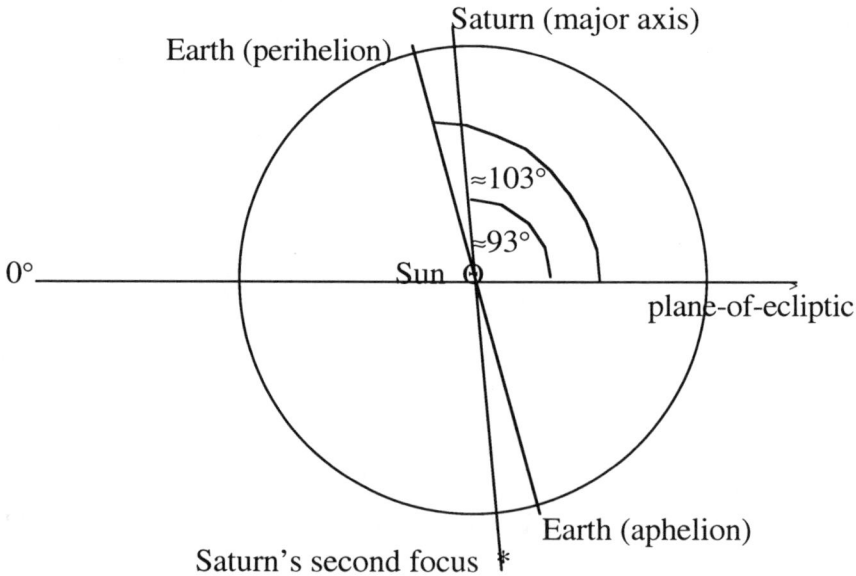

Figure 7.2: Representation of the location of Saturn's second focus on its orbital major axis.

From **Table 7.1** we can predict the location of the second focus for the planet Saturn to be approximately 1.5523E11m from the Sun. This location is outside the path of Earth's orbit and near Earth's aphelion location. The Earth orbital point of closest approach to Saturn's second focus occurs annually near June 23rd (approximately 10 days prior to aphelion and near the summer solstice).

The aphelion separation for Earth is approximately 1.5210E11m, making the orbital separation of this focus approximately 3.16E9m outside Earth's orbit. The approximate separation between Earth orbit and the focus is approximately 1.96 million miles or, approximately 18 gravitational standing wavelengths (wavelength, λ = 1.75E8 m) from Earth.

The Model and an Analysis of the Jovian System

Accurate data for the satellites of a planet becomes difficult at best. Some of the orbits appear nearly circular, having a reported eccentricity of near zero. Data for these orbits will not be included in all of the analyses. Based on the earlier and more extensive analysis of the planetary orbits, following are values for the 'area-per unit-time' calculated using the model and data for the Jovian System:

Satellite	Integer	2A / T E 10 (m²/s)
Andrastea	71	5.7122
Amalthea	84	5.7130
Thebe	93	5.7078
Io	128	5.7085
Europa	166	5.7251
Ganymede	204	5.7042
Callisto	270	5.7221
Leda	647	5.7311
Elara	659	5.7270
Himalia	660	5.7225
Lysithca	669	5.7270
Pasiphae	883	5.7290
Anake	891	5.7297
Sinope	904	5.7312
Carme	917	5.7266

Table 7.9: An analysis using the computer model for '2 A/T' and satellite motion in the Jovian System with an 'error window' of 1.5%. Averages from the search and comparison of the 'twice the area-per-unit-time' for the moons of Jupiter:

- r_1 = 2.65E4 m   • T_1 = .076 s   • λ_1 = 1.67E5 m.

This analysis, $2A / T$, is equivalent to the I_{bar} value for the Jovian System and should be proportional to the mass of the central body, Jupiter's Mass of 1.901E27 kg. The numerical average of this set of data is approximately 5.72E10 m²/s. Following is a calculation using the mass proportionality:

$$I_{bar\,SS} / M_{Sun} = I_{bar\,J} / M_J$$

$$6.04E13 \text{ m}^2/\text{s} / 1.03E21 \text{ kg} = I_{bar\,J} / 9.84E17 \text{ kg, and}$$

$$I_{bar\,J} = 5.74E10 \text{ m}^2/\text{s}$$

Orbital Foci Separation within the Jovian System:

Satellite	Foci Separation (m)	Integer	Separation E 4 (m)
Io	3.4571E6	43	8.4320
Thebe	6.6570E6	83	8.2185
Europa	1.3552E7	169	8.3141
Lysithea	2.5081E9	30021	8.3545
Leda	3.2838E9	39305	8.3547
Himalia	3.6277E9	43423	8.3543
Elara	4.8591E9	58161	8.3546
Anake	7.1656E9	85769	8.3545
Carme	9.3564E9	111993	8.3545
Sinope	1.3035E10	156025	8.3544
Pasiphae	1.7766E10	212653	8.3545

Table 7.10: An analysis using the computer model for foci separation (2c/n) in the Jovian System with an error window of 2.0%. Averages for this search and comparison include:

- $v_1 = 2.19E6$ m/s
- $T_1 = 0.076$ s
- $r_1 = 2.66E4$ m
- $\lambda_1 = 1.67E5$ m.

It would seem the significance of the fine structure constant and its connection to orbits at any scale within the electromagnetic realm may be among the more important concepts associated with this study. Certainly more questions will be identified from an analysis of this work than found answers. But, that is an exciting aspect of good science; the questions always outnumber the reasonable answers! So the search for truth would seem to be unending, an unending need to remain '**Ever Hearing and Ever Seeing.**'

Chapter 7: References and Resources

1. Weast, R. (Editor, 60th ed.), <u>Handbook of Chemistry and Physics</u> , (Boca Raton, Florida: CRC Press, Inc., 1980), p F-175 to F-180. (**note:** Appendix A1.1-A1.2)

2. Wagner, J., <u>Introduction To The Solar System,</u> (Philadelphia: Saunders College Publishing, a division of Holt, Rinehart and Winston, Inc., 1991), p 274.

3. Wagner, J., <u>Introduction To The Solar System,</u> (Philadelphia, Saunders College Publishing, a division of Holt, Rinehart and Winston, Inc., 1991), p A.6-7. (**note:** Appendix A1.1-A1.2)

Chapter Eight: So, ...

Saying it is so, does not make it so. This is an axiom for all facets of life and at every interface of relationship. Science holds that it is possible for an event or process to exist somewhere within the universe, if its existence does not violate the mandates of the scientific, conceptual framework. But, this proposed existence must measure-up to observation and interrogation; if not by first hand evidence, at least by the acquisition of a strong case of circumstantial evidence. Further probing, testing, and refinement must transpire before the event can be declared a scientifically identifiable occurrence.

Prediction and Science

A standard coin has two sides. Often one refers to the two sides of any U.S. coin as having a head and a tail. The flip and landing of this coin presents two nearly equal options for a final state. Mathematicians would argue that each event, the flip and landing, has an equal probability of being either a head or a tail. So, the probability for either a head or tail for any given flip is 50%.

Any given prediction can be either right or wrong. But, unlike the flip of a coin, the probability of a given result being correct may be quite different from fifty percent. This book incorporates two arguments which are couched within the realm of scientific hypothesis: 1) an extended, solar minimum beginning as early as 2001 and 2) an argument for a quantum gravitational model to describe the Solar System. More scientific work is necessary to improve the reliability of both of these hypotheses, clearing the way for the formation of a scientific prediction. In short time the 2001 date will confirm or disprove an earliest beginning for a significant event. Yet, if not in 2001, then it may come within the next two or three, sunspot cycles (multiples of about 11y).

An identifiable periodicity associated with a given event enhances the probability of predicting a reoccurrence for the event. The greater the documentation for the periodicity of a given event, the greater the reliance in being able to predict a next event. When it comes to predicting a next, extended solar minimum, credible evidence for a known periodicity is lacking. The computer model used in Chapter 4 may be a good approximation for the dynamics of the center-of-mass for the Solar System, but the complexity of such an event merits more attention and delineation from both solar scientists and cosmologists.

Therefore, making a prediction for a next, extended solar minimum and proposing a quantum gravitational view for the Solar System are radical ideas at best. So, why does one get involved in such activity? A reasonable response to this question might include: to learn about nature and/or to improve one's understanding of scientific inquiry.

The motivations for learning are diverse and compelling. Some think this type of activity borders on perversion. It certainly is a free-will activity. Why does a hunter hunt? Why does a fisher-person fish? Perhaps it is relaxing, enjoyable, challenging, or a retreat from reality. Who but the participant can say, and then, the response may be given with only a bit of certainty.

For whom will the result prove to be of value? Will the prediction or result move mankind closer to the realm of truth or will it stir the water and muck the view? These are important considerations. For science, progress demands questions and questions require a response. Hopefully the response carries the inquiry process closer to truth, either with a clarification of an erroneous logic or a ratification of logic, reason, and result.

New views for old problems can be an encouragement for others to join in the search. Many hands and/or minds make lighter a weighty task! This exciting aspect of science has propelled the scientific revolution for nearly four centuries. Sharing with one another about a result, an application, or an experiential encounter has been encouraging and exciting.

An Extended Solar Minimum

If this event happens, the results may produce wide scale famines with a potential for plagues. Will our responses to such a challenge be centered on relationship? Will we seek the opportunity to help our neighbor, joining in mutual encouragement and survival?

Knowing how others before us managed to survive an extended, drought enhances our options; it provides encouragement for a civil response to a dire situation. Problems always bring opportunities, new views to incorporate in making both personal and corporate decisions.

Predicting an event associated with a physical phenomenon is risky. But even more risky is the challenge of predicting human response to a forthcoming event and its effects.

A Model for Gravitational Waves within the Solar System

This model is an interesting thread of logic. If any consistencies between nature and the model exist, additional theoretical and experimental work will clarify the descriptions and interpretations for gravitational interactions. These interactions may be fundamental to the existence of living systems.

Appendix A1.1

Common data for the Solar System

Planet	Mass (kg)	Avg. Sep. (m)	Err (+/-) (m)	Aphelion Sep. (m)	Err (+/-) (m)
Mercury	3.181E23	5.795E10	1.3E8	6.986E10	1.6E8
Venus	4.883E24	1.0811E11	1.0E8	1.0885E11	1.0E8
Earth	5.979E24	1.4957E11	7.0E7	1.5207E11	7.0E7
Mars	6.418E23	2.2784E11	2.2E8	2.4912E11	2.0E8
Jupiter	1.901E27	7.7814E11	1.7E8	8.1580E11	2.4E8
Saturn	5.684E26	1.4270E12	9.0E8	1.5045E12	3.1E8
Uranus	8.682E25	2.8703E12	1.5E9	3.0023E12	4.6E9
Neptune	1.027E26	4.4999E12	3.8E9	4.5368E12	5.0E9
Pluto	1.08E24	5.909E12	1.9E10	7.375E12	2.4E10

Planet	Perihelion Sep. (m)	Err (+/-) (m)	Eccentricity	Err (+/-)
Mercury	4.604E10	1.0E8	0.2056	0.0002
Venus	1.0737E11	1.0E8	0.0068	0.0001
Earth	1.47076E11	7.0E7	0.0167	0.0001
Mars	2.0656E11	2.0E8	0.0934	0.0002
Jupiter	7.4048E11	2.2E8	0.0484	0.0002
Saturn	1.3495E12	3.1E9	0.0543	0.0021
Uranus	2.7383E12	4.5E9	0.0460	0.0015
Neptune	4.4630E12	4.9E9	0.0082	0.0007
Pluto	4.443E12	1.5E10	0.2481	0.0006

Appendix A1.2

Planet	*Longitude Perihelion	Aphelion speed(m/s)	Perihelion speed(m/s)	period (seconds)
Mercury	77.144	3.8824E4	5.8921E4	7.60234E6
Venus	131.290	3.4780E4	3.5256E4	1.94141E7
Earth	102.596	2.9278E4	3.0272E4	3.155815E7
Mars	335.691	2.1964E4	2.6490E4	5.93553E7
Jupiter	14.010	1.2435E4	1.3700E4	3.74320E8
Saturn	92.665	9.1284E3	1.0177E4	9.29604E8
Uranus	172.736	6.4902E3	7.1161E3	2.65114E9
Neptune	47.867	5.3833E3	5.4723E3	5.20027E9
Pluto	222.972	3.6763E3	6.1024E3	7.83735E9

Body	Mass (kg)	Radius (m)	Avg. Sep (m)	Eccentricity
*Sun	1.989E30	6.960E8	- - -	- - -
*Moon	7.349E22	1.738E6	3.844E8	0.0549

1. Weast, R. (Editor, 60th ed.), <u>CRC Handbook of Chemistry and Physics</u>, (Baca Raton, Florida: CRC Press, Inc., 1980), p F176-F180.

*2. Wagner, J., <u>Intoduction To The Solar System</u>. (Philadelphia: Saunders College Publishing, 1991), Appendix 2-3.

Appendix A2.1: About the Ellipse and Hyperbola

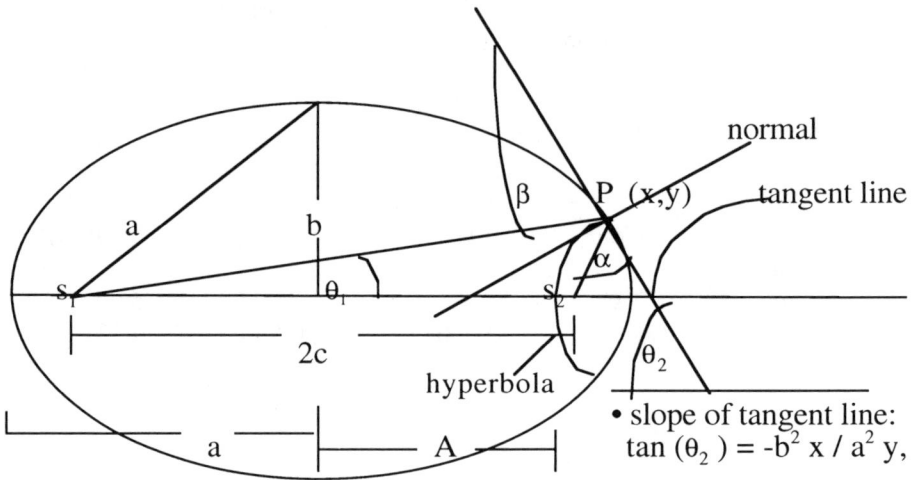

normal

tangent line

β P (x,y)

α

a b

S₁ θ_1 S₂

2c

hyperbola

θ_2

• slope of tangent line:
$\tan(\theta_2) = -b^2 x / a^2 y$,

a A

• angle α = angle β, and the angle formed by PS_1 and the normal equals the angle formed by PS_2 and the normal.

Figure A2.1: Parametric connections between an ellipse and a hyperbola.

About An Ellipse:	**About A Hyperbola:**
• $x^2/a^2 + y^2/b^2 = 1$	• $x^2/A^2 + y^2/B^2 = 1$
• $e = c/a$	• $e = c/A$
• $x = r_1 \cos \vartheta$ - ae	• $r_1 - r_2 = 2A = n\lambda$
• $y = r_1 \sin \vartheta$	
• $r_1 + r_2 = 2a$	

1) $r_2 = 2a - r_1$ and $r_2^2 = (2a - r_1)^2 = 4a^2 - 4a\,r_1 + r_1^2$

2) $r_2^2 = r_1^2 + 4\,a^2\,e^2 - 4r_1\,ae\,\cos\vartheta$ [law of cosine]

    $\therefore\ 4a^2 - 4a\,r_1 + r_1^2 = r_1^2 + 4a^2\,e^2 - 4r_1\,ae\,\cos\vartheta$

     $4a^2 - 4\,a^2\,e^2 = 4a\,r_1 - 4a^2 e^2 - 4r_1\,ae\,\cos\vartheta$ and

   a. $r_1 = a[1-e^2]/[1 - e\cos\vartheta]$ with, $b^2 = a^2 - c^2 = a^2(1-e^2)$

   b. $r_1 = [b^2/a][1/(1 - e\cos\vartheta)]$ where, $[b^2/a]$ is one-half the ellipse latus rectum, [a] depends on the total orbital energy, and [b] depends the orbital angular momentum, and

   c. $r_1 = a + xe$

Appendix A2.2: Parametric Connection (part I)

3) at common intercept point (ellipse and hyperbola)

$$r_1 + r_2 = 2a \qquad \text{and} \qquad r_1 - r_2 = 2A = n\lambda$$

$$2 r_1 = 2 (a+A) \quad \text{with}$$

a. $r_1 = a + A$ and

b. $r_2 = a - A$

(1) $A = r_1 - a = \{b^2 / a\} \{[1 / (1 - e \cos\vartheta)] - a \}$

(2) $A = (a + xe) - a = xe$ or,

$x = A / e$ [used within the computer model]

(3) $y^2 = r_1^2 - (x+c)^2 = b^2 - x^2 (1 - e^2)$

Appendix A2.3: Parametric Connection (Part II)

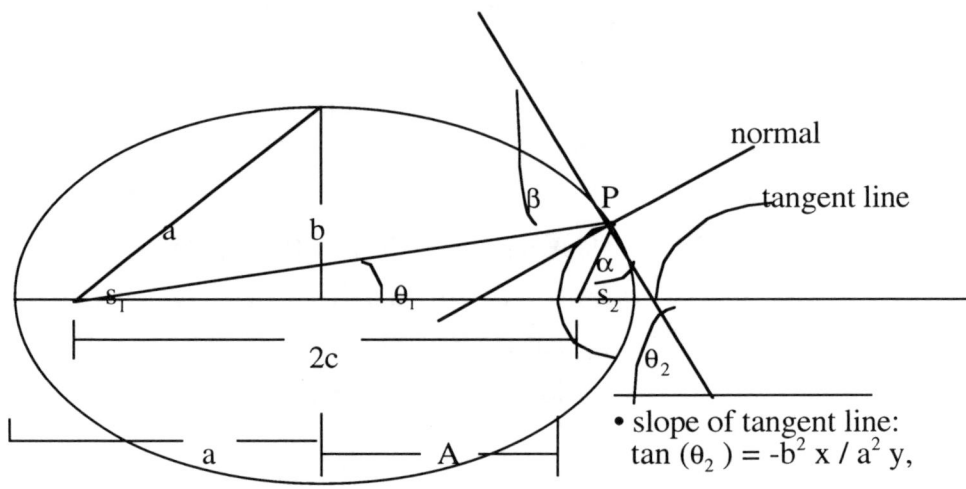

normal

tangent line

• slope of tangent line:
$\tan(\theta_2) = -b^2 x / a^2 y$,

• angle α = angle β, and the angle formed by PS_1 and the normal equals the angle formed by PS_2 and the normal.

Figure A2.3: Tangency characteristics for a point on an ellipse.

About An Ellipse
 • $PS_1 + PS_2 = 2A$
 • $a^2 y^2 = a^2 b^2 - b^2 x^2$

[differentiating both sides]

 • slope of tangent line

   1) $2a^2 y\, dy = -2b^2 x\, dx$

   2) $dy / dx = -b^2 x / a^2 y$

   3) $\tan \theta_2 = |-b^2 x / a^2 y| = [b^2 / a^2][x / y] = dy / dx$

   $\quad b^2 = a^2(1 - e^2)$ with $e = c / a$

   $\quad b^2 / a^2 = (1 - e^2)$ and,

   4) $dy / dx = [x / y][e^2 - 1]$

   5) $\tan \theta_2 = [1 - e^2][x / y]$

 • slope of normal to tangent line

   $-1 / [dy / dx] = -dx / dy = a^2 y / b^2 x = [1 / (1 - e^2)][y / x]$

Appendix A2.4: Parametric Connection (Part II)
About A Hyperbola

- $PS_1 - PS_2 = 2A = n\lambda$
- $A^2 y^2 = B^2 x^2 - A^2 b^2$

[differentiating both sides:]

- slope of tangent line

    1) $2A^2 y\, dy = 2B^2 x\, dx$ with, $e_h = c / A$

    2) $dy / dx = [B^2 x] / [A^2 y]$ and $[B^2 / A^2] = [e_h^2 - 1]$

- slope of normal to tangent line

    $- 1 / [dy / dx] = - dx / dy = -A^2 y / B^2 x = -[1 / (e_h^2 - 1)] [x / y]$

Parametric Connection: At Common Intercept

- $y_e^2 = b^2 / a^2 [a^2 - x^2] = [1 - e_e^2] [a^2 - x^2]$
- $y_h^2 = B^2 / A^2 [x^2 - A^2] = [e_h^2 - 1] [x^2 - A^2]$

    1) $[1 - e_e^2] [a^2 - x^2] = [e_h^2 - 1] [x^2 - A^2]$

        $a^2 - a^2 e_e^2 - x^2 + x^2 e_e^2 = x^2 e_h^2 - x^2 - A^2 e_h^2 + A^2$

        $a^2 [1 - e_e^2] + x^2 e_e^2 = A^2 [1 - e_h^2] + x^2 e_h^2$

        $a^2 [1 - c^2 / a^2] + x^2 c^2 / a^2 = A^2 [1 - c^2 / A^2] + x^2 c^2 / A^2$

        $[a^2 - c^2] + x^2 c^2 / a^2 = [A^2 - c^2] + x^2 c^2 / A^2$

        $[a^2 - A^2] = x^2 c^2 [a^2 - A^2] / a^2 A^2$

        $x^2 c^2 = a^2 A^2$ or,

        $A^2 = x^2 c^2 / a^2$ and $A = x e_e$:

        a. $x = A / e_e$ with **A** being proportional to **x** and e_e being the 'rate of change', and

        b. $y^2 = b^2 - x^2 [1 - e_e^2]$

    2) Is the normal to the tangent line of an ellipse equivalent to the tangent line for the hyperbola intersecting at the point of tangency on the ellipse?

        $[1 / (1 - e_e^2)] [y / x] = [B^2 x] / [A^2 y]$ with $A^2 = x^2 e_e^2$

        and $y^2 = (1 - e_e^2) (a^2 - x^2)$

        $B^2 = [A^2 y^2] / [(1 - e_e^2) x^2] = [e_e^2 y^2] / [1 - e_e^2] = e_e^2 (a^2 - x^2)$ and

        $B^2 = c^2 - A^2$ [proof by induction: an equation for a hyperbola]

Personal Notes

Personal Notes

About the Author:

David Reid earned a B.Sc. in Physical Science from Kansas State University and a M.A. in Combined Science: physics and chemistry (NSF) from the University of Mississippi. For thirty-four years he taught physical science to secondary science students in the Jefferson County Public Schools; Golden, Colorado. Physics education was instrumental in shaping his view of science. His instructional assignments included working with implementation teams in 'Jeffco' on the following projects:

- Introductory Physical Science,
- Physical Science II, and
- Use of the Computer in Physics and Mathematics (NSF).

David was an instructor in numerous, teacher workshops during the span of his teaching career. These workshop experiences involved instruction of physical science, physics, and/or the use of the computer in classroom instruction. He was a 'Physics Teacher Resource Agent' (AAPT) and a member of the Labnet Team at Tufts University and the Technical Education Resource Center (TERC).

Order Form

If you would like to order additional copies of **'Ever Hearing...and Ever Seeing...'**, please complete the following order form.

_____

Name: _____

Address: _____

_____ Zip _____

Phone: (   ) _____-_____

Costs:

1. less than ten copies: ($11.95)   ...   11.95 X _____ = _____

2. less than twenty copies: $11.95 X _____ = _____ - 10% = _____

3. more than twenty copies: $11.95 X _____ = _____ - 20% = _____

   • tax rate (6.3%)

   • handling/mailing (5%)

Summary of Cost:

Books: _____

Tax: _____

Handling: _____

Total: _____

Thank you!

Plumbline Publishing and Software
13287 W. Montana Place
Lakewood, Colorado 80228